Maria Luisa Chiofalo

Screening effects in bipolaron theory and high-temperature superconductivity

TESI DI PERFEZIONAMENTO

SCUOLA NORMALE SUPERIORE
PISA - 1997

Tesi di perfezionamento in Fisica sostenuta il 12 ottobre 1996

ISBN: 978-88-7642-279-9

Contents

Chapter 1

Introduction

The subject of the present thesis is a study on bipolaron formation in heavily doped and polar materials. The study is applied to the physics of superconducting compounds with high transition temperature (T_c), which are characterized by a short coherence length and a low carrier density with respect to conventional metallic superconductors. In particular this thesis addresses the problem of the formation and superconducting properties of real space electron (hole) pairs in these systems and their dependence on the carrier density. The pairing mechanism results from the dynamical cooperation of phononic and electronic degrees of freedom which characterize the heavily doped polar material. We call these pairs bipolarons (BP) or better *biplasmapolarons* (BPP) because they can be thought as quasi-particles composed by two electrons (holes) and their cloud of virtual phonons and plasmons. The real space pairs have an average radius of the order of the coherence length in high-T_c superconductors (high-T_cSC) and are expected to Bose condense below a given critical temperature.

While the phenomenology of high-T_c materials is discussed in more detail in the next chapter, we wish to sketch now the physics we have in mind. We consider a cubic material. It is very polarizable and the longitudinal optical (LO) phonons are mainly responsible for the electron-phonon coupling. The introduction of extra charges upon doping the material leads to correlation effects and to the development of a coupling between LO phonons and electrons. Because of the large values of the static dielectric constant, the effective Bohr radius a_0 of the electron system turns out to be large; it follows that, in spite of the low carrier density and whence of the large average interparticle distance r_0, the many-particle system is characterized by an adimensional coupling strength $r_s = r_0/a_0$ of the order or even smaller than 1. Therefore we may consider the system to be in a weakly interacting regime, where the Random Phase Approximation can be applied [Pines63].

The study of the system is based on a microscopical hamiltonian approach. We consider two electrons (or holes) singled out from the conduction (valence)

1

band; they directly repel each other and interact with the whole electron gas exchanging long range plasma excitations and with the ions exchanging optical phonons. The plasmon-LO phonon coupling is also considered. The electron-phonon interaction is treated within the Fröhlich scheme of polaron theory [Froh50,Froh63]; a similar scheme turns out to be appropriate for the description of the electron-plasmon interaction. A reasonable evaluation of the adimensional Fröhlich electron-phonon coupling constant α [Froh63] for high-T_c cuprates is $\alpha \sim 8$. Such a value of α is near to the upper (lower) limit for which the intermediate (strong) coupling polaron theory applies. We stay within the intermediate coupling regime and thus we use the variational Lee, Low and Pines approach [LLP53].

The pioneering work on large bipolaron formation within the intermediate electron-phonon coupling regime is due to Bassani et al. [Bassani91] and Devreese et al. [Devreese91]. The stability of a single large bipolaron (without the electron gas) was found below a critical value η_c of the material ionicity; η is the ratio of the high-to-low frequency dielectric constants and it is smaller the larger is the ionicity. As far as $\eta < \eta_c$ the bipolaron exists as a bound state. The value of η_c depends on the electron-phonon coupling constant α: the larger α, the larger η_c can be. A further result concerns the stability of the bipolaron with higher orbital angular momenta L and it is found that bipolarons are more stable in s-wave pairing rather than in higher angular momentum-wave. On the other hand, it is found that the bipolaron is more deeply bound in two rather than in three dimensions. It is noteworthy that the stability region in the space of the introduced parameters obtained by Bassani et al. with their variational calculation is in excellent agreement with the conclusions independently reached by Devreese and coworkers [Devreese91] within a path integral calculation.

The first results concerning the binding energy of the two electrons interacting not only with phonons but also with plasmons were presented by Iadonisi and coworkers [Cataud91,Cataud92]. The idea was that, at low electronic densities, the conduction electrons partecipate to the binding energy of the bipolaron not only because they can screen the electrostatic interactions, but also because they lead to a cooperative effect with the phonons, the two excitations being treated on the same footing. Such a plasmon mechanism has been considered independently also by other authors [Gersten88,Kresin88].

In the present work we extend the theory to the calculation of both BPP binding energy and effective mass [Iadonisi93,Chiofalo92]. Then we discuss the features of the effective electron-electron interaction. We find that in the low density range BPPs are truly bound pairs, their binding energy being a decreasing function of the density as the typical densities in high-T_cSC are approached. A similar behaviour is displayed by the BPP effective mass. At larger density the effective electron-electron potential retains an attractive well even though the BPP energy has become positive (the BPP is unbound). This suggests the existence of a resonant BPP state with a finite lifetime and the coexistence of bipolarons and polarons in the system. This assumption is used to calculate a number of density and temperature dependent superconducting properties [Iadonisi95] within a phenomenological boson-fermion model. The predicted results are then compared with experimental data and a satisfactory agreement is obtained.

The thesis is organized as follows. Chapter 2 is devoted to a description of the experimental scenario in high-T_c superconductivity. An effort is made to give an idea of the main results concerning the normal as well as the superconducting-state properties of high-T_c cuprates, whether controversial or well established. Particular emphasis is given to those aspects which are relevant for the thesis. Thus the signatures of lattice and polaron effects in the superconducting mechanism of high-T_c cuprates are discussed. In addition references to some of the most cited theoretical approaches are given. The chapter ends up with a discussion on the collocation of the present work with respect to existing theoretical models.

In chapter 3 the derivation of the BPP hamiltonian is presented. Since the part of the hamiltonian concerning the boson (plasmon and phonon) fields is exactly diagonalized, the system is reduced to two charges interacting with two independent renormalized fields. It is also shown that the results concerning the interaction of the external charges with the renormalized fields can be obtained through a dielectric formulation of the problem. The corresponding total dielectric function contains the sum of two contributions, the ionic part and the plasmonic one and corresponds to a Random Phase Approximation for the interactions involved [Mahan].

The description of the variational procedure used to solve the hamiltonian is contained in Chapter 4. The use of coherent states for the description of phonons and plasmons is described.

Chapter 5 is devoted to the discussion of the effective electron-electron interaction and to some comments concerning the large polaron approach. A brief discussion on the existence of resonant bipolaron states is given [Cataud96].

Chapter 6 contains the numerical results concerning the microscopical model. The BPP binding energy and the effective mass are calculated within a self-consistent variational procedure. All the calculations are done self-consistently up to n$\simeq 10^{20}$ cm^{-3}. In this limit $nR_b^3 \leq 1$ (R_b is the bipolaron radius) and therefore it is meaningful to consider only one pair of electrons in interaction with the others [Iadonisi93,Chiofalo92].

In Chapter 7 the thermodynamics of the coexistence of bosons and fermions within the boson-fermion model is presented. Some superconducting properties are calculated: the critical temperature as a function of the total carrier density and as a function of the ratio between the condensate density and the BPP effective mass; the condensate fraction and chemical potential as a function of the temperature and of the total carrier density. The comparison with experimental data is discussed [Iadonisi95].

Finally in chapter 8 we draw our conclusions on the relevance of such a physics to the high-temperature superconductivity.

Given the fundamental role of the screening effects in the system under consideration, some attention has been devoted to the study of the properties of a charged boson fluid on more general grounds; in particular static local field theories for the density-density response function as well as a sum rule approach have been used [Chiofalo94,Conti94], [Chiofalo95,Chiofalo96]. The main results are summarized in Appendix B.

Chapter 2

A review of relevant experiments on high-T_C superconductivity

2.1 Introduction

The liquefaction of liquid Helium by Kamerlingh Onnes at the beginning of this century has been a revolutionary event for the low temperature physics. The investigation of the properties of some simple metals cooled down to such low temperatures leads to a remarkable discovery: their resistivity goes to zero below a well defined (and finite) temperature. Later on another striking feature of the superconducting materials has been shown by Meissner and Ochsenfeld [MeissOchs33]: when subjected to a magnetic field and then cooled below T_c (or viceversa), they expel the field lines.

The leading theoretical idea to explain such peculiar features is the formation of pairs and is nicely expressed in Gamow's limerick [Blatt64] on Ogg's *bielectron* theory [Ogg46]

In Ogg's theory it was his intent
That the current keep flowing, once sent;
So to save himself trouble,
He put them in double,
And instead of stopping, it went.
Gamow.

In the Fifties Landau and Ginzburg [GinzLandau50] faced the problem of superconductivity within a phenomenological approach, characterizing the appeerence of the superconducting state through the presence of a macroscopic quantum wave function of the condensate system. Blatt, Butler and Schafroth [Blatt64,Scha55,Scha57] gave a semiphenomenological description within a two fluid model; they postulated the coexistence of a resonant state of two electrons, statistically treated as pointlike bosons of charge $2e$, and

Maria Luisa Chiofalo

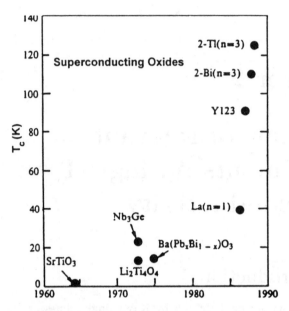

Figure 2.1: Superconductor critical temperatures have grown up by an order of magnitude during the last ten years.

unpaired electrons. However the experimentally found critical temperatures implied unrealistic carrier densities ($\simeq 10^{17}$ cm^{-3}) and pair masses ($\simeq 10^{8}$ electron masses). It took 40 years after Onnes' discovery before a fully satisfactory microscopical model for superconductivity was put forward by Bardeen, Cooper and Schrieffer [BCS57].

The Bardeen-Cooper-Schrieffer (BCS) theory represents the standard theoretical framework for superconductivity. BCS theory allows the treatment of a fermion fluid with an effective attractive interaction which leads to the formation of a condensate of spatially overlapping pairs. The successful application of the theory to metal and alloy superconductors (the so called "conventional"-low temperature- superconductors) is based on a phonon-mediated microscopic mechanism which is responsible for pairing in momentum space.

Among the conventional materials, it appeared that the poorer was the normal state conductivity the higher the transition temperature, the best being Nb_3Ge with $T_c = 23.2°K$. On these grounds, it has been natural to pursue the search for higher critical temperature compounds among materials with stronger lattice polarizability and then enhanced electron-phonon coupling. Historically, this was the leading idea [BednMuller88] of Georg Bednorz and Alex Müller when they found [BednMuller86] in 1986 a new kind of material with $T_c = 30°K$. The material was $(La_{2-x}M_x)CuO_4$ (with M=Sr, Ba or Ca), the first of a new class of superconductors, broadly referred to as cuprates or cuprous oxides. In a few years the critical temperatures of superconducting

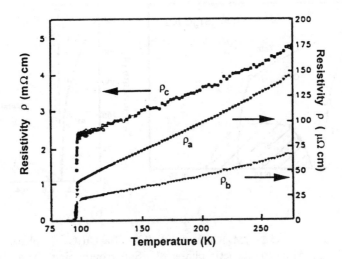

Figure 2.2: Resistivity ρ of YBa$_2$Cu$_3$O$_{7-\delta}$ as a function of temeperature T measured along the three crystallographic directions of its orthorhombic structure (see Sec. 2.2).

materials grew up to 135 $^\circ K$ (Fig. 2.1), well above the boiling point of liquid nitrogen.

The discovery of high-T$_c$ superconductors (high-T$_c$SC) represents a new challenge for both experimental and theoretical physics. We report in Fig. 2.2 an example of the resistivity curve of one of the new materials, namely YBa$_2$-Cu$_3$O$_{7-\delta}$ (see below). Concerning the cuprates, they are very complex indeed and have quite unusual properties in the superconducting as well as in the normal state. Fig. 2.3 shows the phase diagram of two among the most well known cuprates, namely (La$_{2-x}$M$_x$)CuO$_4$ and YBa$_2$Cu$_3$O$_{7-\delta}$. The phase diagram shows how rich is the physics of these new materials, running from an antiferromagnetic insulator state to a metallic state, to superconductivity, to structural transitions between tetragonal and orthorhombic phases. It evidences also how crucial is doping in this variety of different physical states.

On the experimental side many technical problems are to be faced in the investigation of the high-T$_c$ compounds, in order to disentangle the features which are relevant to understand the underlying physics. Cuprates are strongly anisotropic and made of planes of copper and oxygen separated by planes of other oxides and rare earth elements. In addition, the structure of the synthesized samples can be twinned along the axis perpendicular to the CuO planes. A further complication stems from the non-stoichiometric nature of these materials, which is crucial in order to span the whole phase diagram. This

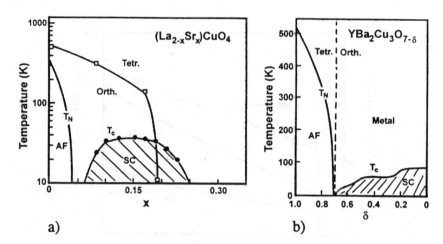

Figure 2.3: (a) $(La_{2-x}M_x)CuO_4$ and (b) $YBa_2Cu_3O_{7-\delta}$ phase diagrams. AF=Insulating Antiferromagnetic phase; SC=Superconducting phase; T_N = Néel temperature; T_c = Superconducting critical temperature.

is obtained doping the insulating material by the addition of elements like Ba, Sr or Ca in the case of La_2CuO_4, or acting on the amount of oxygen or both possibilities. The doping introduces additional charges (holes in most cases), the number of which strongly depends on the material processing. It follows that the same type of measurements on different samples can give different and sometimes contradictory data. At the very beginning further difficulties were also due to the polycrystalline nature of the samples, which fortunately have been now overcome by the availability of good single crystals.

From a theoretical point of view, high-T_c superconductivity has attracted people with different backgrounds and technical languages. Looking once again at the phase diagram in Fig. 2.3 it is not surprising that many different microscopical pairing mechanisms have been proposed to explain superconductivity in cuprates. In fact, the excitations that could lead to an effective attractive interaction may have different origin, either phononic, magnetic, or pure electronic correlation effects or even a cooperation of two or more of them. The point is that whatever the mechanism is, it must be sufficiently strong in order to account for the observed short coherence length of the new materials with respect to the conventional ones. Beyond the "choice" of the mechanism, a more fundamental and in some sense technical question arises: whether BCS theory has to provide the many-particle tools to face the problem or Bose-Einstein Condensation (BEC) of real space tightly bound pairs would be more appropriate. Along these lines, large efforts have been made trying a phenomenological description of the crossover from BCS theory to BEC.

After almost ten years, the questions are many and still too difficult to be answered to. However some definite conclusions can be drawn, at least from the experimental viewpoint. The next sections contain a brief phenomenological description of high-T_cSC cuprates.

2.2 Structure

Let us start with a little presentation of the crystal structure of cuprates together with some needed nomenclature, which will be useful in the following.

The crystal structure of cuprates is tetragonal (body-centered -bct- or primitive -pt- tetragonal) in most cases, possibly with small orthorhombic distorsions. Cuprates can be thought as made of a given number of adjacent CuO planes parallel to the a-b plane of the tetragonal structure; the planes are separated by spacing layers of different composition. From this viewpoint, cuprates can be classified according to this number of CuO planes per formula unit. In addition, adjacent CuO planes sandwich a layer of alkali or rare earth atoms. This is the general recipe.

Tabs.2.1 and 2.2 lists some of the so far synthesized high-T_cSC , together with their structural characteristics. As far as the notation is concerned, there are many possibilities in the literature; here it is considered the one which is made by the symbol of the metal-based compound followed by four numbers; each of the first three represents the chemical composition with respect to all the component atoms but copper and oxygen; the last one indicates the number of the adjacent CuO planes. For instance $Tl_2Ba_2Ca_2Cu_3O_{10}$ is the thallium based compound with 2 thallium, 2 barium and 2 calcium atoms in the chemical formula; in addition it has three adjacent CuO planes and therefore it is represented by the notation Tl2223. Yttrium-based compounds constitute an exception: only three numbers are present and they refer to yttrium, barium and copper chemical composition. In addition the lanthanum-based compound is traditionally simply referred to as 214. From the Table it appears also that the higher is the number of CuO planes, the higher the transition temperature.

Figs.2.4,2.5 display the lattice structure and the unit cell for some of the listed compounds. The following comments are in order:
– La-, Tl2-, Bi- and Nd-based compounds have a body-centered tetragonal structure, which contains two formula units per conventional unit cell; Tl1-based compounds have primitive tetragonal unit cells; Y-based are orthorhombic.
– La- and Y-based compounds have structural phase transitions between orthorhombic and tetragonal structures. However in the case of YBa_2Cu_3-$O_{7-\delta}$ the "transition" is not at constant doping. This is shown in Fig. 2.3a-b.
– Tab. 2.2 displays the composition of the isolation planes. In the case of $YBa_2Cu_3O_{7-\delta}$ they are made of Cu and O atoms (along the b axis in Fig. 2.5)

General formula	Compound	T_c (max) (K)	n	Notation
$(La_{2-x}Sr_x)CuO_4$	$(La_{2-x}Sr_x)CuO_4$	38	1	214
$Tl_2Ba_2Ca_{n-1}Cu_nO_{4+2n}$	$Tl_2Ba_2CuO_6$	80	1	Tl2201
	$Tl_2Ba_2CaCu_2O_8$	108	2	Tl2212
	$Tl_2Ba_2Ca_2Cu_3O_{10}$	125	3	Tl2223
$Bi_2Sr_2Ca_{n-1}Cu_nO_{4+2n}$	$Bi_2Sr_2CuO_6$	20	1	Bi2201
	$Bi_2Sr_2CaCu_2O_8$	85	2	Bi2212
	$Bi_2Sr_2Ca_2Cu_3O_{10}$	110	3	Bi2223
$(Nd_{2-x}Ce_x)CuO_4$	$(Nd_{2-x}Ce_x)CuO_4$	30	1	–
$TlBa_2Ca_{n-1}Cu_nO_{3+2n}$	$TlBa_2CuO_5$	50	1	Tl1201
	$TlBa_2CaCu_2O_7$	80	2	Tl1212
	$TlBa_2Ca_2Cu_3O_9$	110	3	Tl1223
	$TlBa_2Ca_3Cu_4O_{11}$	122	4	Tl1234
$YBa_2Cu_3O_{7-\delta}$	$YBa_2Cu_3O_7$	92	2	Y123
$YBa_2Cu_4O_8$	$YBa_2Cu_4O_8$	80	2	Y124
$Y_2Ba_4Cu_7O_{14+x}$	$Y_2Ba_4Cu_7 O_{14}$	40	2	Y247

Table 2.1: List of some high-T_cSC compounds. From left to right: the general formula, the compound, the critical temperature, the number n of CuO planes, the notation for the compound.

Compound	Isolation layers	Layers between two CuO planes
$(La_{2-x}Sr_x)CuO_4$	LaO	–
$Tl_2Ba_2CuO_6$	$BaO + 2TlO + BaO$	–
$Tl_2Ba_2CaCu_2O_8$	$BaO + 2TlO + BaO$	Ca
$Tl_2Ba_2Ca_2Cu_3O_{10}$	$BaO + 2TlO + BaO$	Ca
$Bi_2Sr_2CuO_6$	$SrO + 2TlO + BaO$	–
$Bi_2Sr_2CaCu_2O_8$	$SrO + 2TlO + BaO$	Ca
$Bi_2Sr_2Ca_2Cu_3O_{10}$	$SrO + 2TlO + BaO$	Ca
$(Nd_{2-x}Ce_x)CuO_4$	NdO	–
$TlBa_2CuO_5$	$BaO + TlO + BaO$	–
$TlBa_2CaCu_2O_7$	$BaO + TlO + BaO$	Ca
$TlBa_2Ca_2Cu_3O_9$	$BaO + TlO + BaO$	Ca
$TlBa_2Ca_3Cu_4O_{11}$	$BaO + TlO + BaO$	Ca
$YBa_2Cu_3O_7$	CuO	Y
$YBa_2Cu_4O_8$	$2CuO$	Y
$Y_2Ba_4Cu_7O_{14}$	$2CuO$	Y

Table 2.2: From left to right: the compound formula, the composition of the isolation layers (sometimes more than one) and the composition of the layers between the adjacent CuO planes.

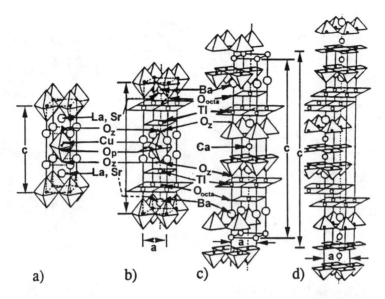

Figure 2.4: The lattice of different compounds: (a) 214; (b) Tl2201; (c) Tl2212; (d) Tl2223. The lattice structure is body-centered tetragonal and contains *two* formula units.

and referred to as "chains".
– There are different kind of oxygen atoms. Following the notation from Ref. [Burns92], one may distinguish: oxygens in the CuO plane (O_p) along both the a and the b direction, four of them surrounding a Cu atom; apical oxygens (O_{ap}) along the c direction, being just above or below the Cu atoms in the CuO planes; octahedrally coordinated oxygens (O_{oct}), belonging to the isolation metal-oxyde planes; chain oxygens (O_c) along the b direction in the case of YBa$_2$Cu$_3$O$_{7-\delta}$.
– As far as the doping is concerned, Tab. 2.1 indicates the kind of substitutions which introduce extra charges. In all cases but the Nd-based compound the charges are holes. In YBa$_2$Cu$_3$O$_{7-\delta}$ the holes can be introduced by oxygen doping. Fig. 2.5b shows that in YBa$_2$Cu$_3$O$_6$ the O atoms are removed from the chains and the lattice structure is now tetragonal.
– Typical Cu-O bond lengths in CuO planes are in the range 1.9-1.94 Å, while Cu-O_{ap} distances are of the order of 2.4 Å. Therefore copper and oxygen in the CuO planes are strongly covalently bonded, whereas the apical oxygen is very weakly bonded to the Cu atoms in the plane. On the contrary, in YBa$_2$Cu$_3$O$_{7-\delta}$ O_{ap} atoms are strongly bonded to the copper in the chains.

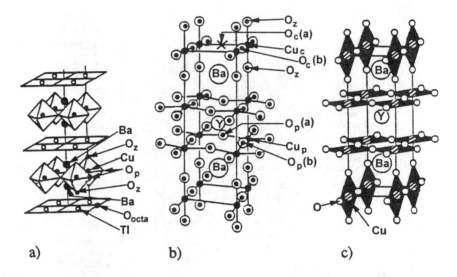

Figure 2.5: The lattice of different compounds: (a) Tl1201; (b) Y123 (O$_6$); (c) Y123. The lattice structure is primitive tetragonal in (a) and (b), orthorhombic in (c). The structure contains *one* formula unit and it is shown doubled for comparison with Fig. 2.4.

2.3   Universal behaviour in Uemura's plot

Universal behaviours of given physical quantities have been extensively looked for in high-T$_c$SC in order to illuminate on the choice of a theoretical framework to tackle the problem. The famous Uemura's plot [Uemura91,Uemura91b,Uemura93] (Fig. 2.6) is a log-log plot where the measured critical temperatures are displayed as a function of the Fermi temperature for a number of superconducting compounds cuprates, alkali-doped C$_{60}$, the so called conventional and strong coupling superconductors and others which have not been mentioned so far (BKBO, heavy fermions, BEDT and Chevrel phases) [Chevrel].

The critical temperatures are measured from muon-spin relaxation experiments (μSR), which will be described later in Sec. 2.4.1. μSR allow an indirect measurement of the ratio n_s/m^* of the superfluid density n_s at $T = 0$ to the superconducting carrier mass m^*. The previous measurement, combined with the observed values of the Sommerfeld constant $\gamma \propto n_s^{1/3}m^*$, leads to an estimate of the Fermi energy $E_F = (\hbar^2/2)(3\pi^2)^{2/3}n_s^{2/3}/m^*$ at $T = 0$ and whence of the Fermi temperature $T_F = E_F/k_B$. This is indicated by the full line $T = T_F$ in the log-log graph. The dashed line represents the transition temperature of an ideal three-dimensional Bose gas, which is given by $T_B = (1.04\hbar^2)n_s^{2/3}/m^*$ considering a density $n_s/2$ and a mass $2m^*$.

When the different high-T$_c$ superconducting materials are shown in the plot, it appears evident that they belong very clearly to a common line close

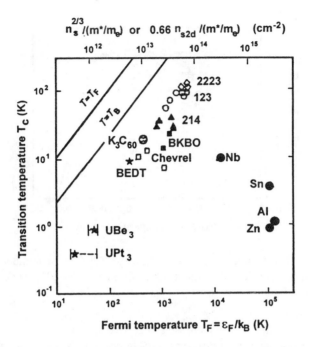

Figure 2.6: Log-log plot of the transition temperature T_c against the Fermi temperature T_F. The Fermi temperature line $T = T_F$ and the critical temperature $T = T_B$ for ideal 3D Bose-Einstein condensation are also displayed (after Ref. [Uemura91]). Note that all the *unconventional* superconductors share a common line almost parallel to T_F and T_B, which are $\propto n_s^{2/3}/m^*$ (see text).

to $T = T_B$, whereas conventional (Sn, Al, Zn in the figure) and strong coupling (Nb) superconductors lie very far away in the graph.

It is clear that Uemura's plot does not tell us anything about the microscopical mechanism underlying the non conventional superconductors. In particular, the compounds sharing the same universal line could have different microscopical mechanisms. Nevertheless, a definite conclusion can be drawn: they lie somewhere in between pure BEC and standard BCS superconductors. This kind of behaviour is also found in the chemical potential as a function of the temperature, which is discussed in Sec. 2.4.6.

2.4   Superconducting-state properties

2.4.1   Critical temperature

A universal relationship between the critical temperature and the hole density in cuprates has been pointed out by Zhang and Sato [ZhSato93] and is shown

Maria Luisa Chiofalo

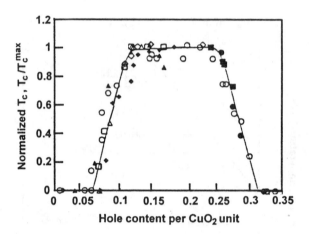

Figure 2.7: Critical temperature normalized to T_c^{max} for a number of cuprates plotted as a function of the hole content per CuO$_2$ unit [ZhSato93]. Different symbols refer to different families.

in Fig. 2.7. Plotting for every cuprate the critical temperature normalized to its maximum one as a function of the hole content, it is found that all the compounds share the same universal curve.

Muon Spin Relaxation experiments

Muon-spin-relaxation experiments have been extensively used to measure the correlation between T_c and the London penetration depth λ (see below) and therefore the ratio of the superconducting density n_s to the superconducting carrier mass m^* [Uemura91,Uemura91b]. Muon spins precess around an applied external magnetic field (with a value in between the two critical magnetic fields); their depolarization is due to the distribution of local magnetic fields in the vortex state (see below) and the measured relaxation rate σ is proportional to the width of such a distribution. On the other hand the distribution width is proportional to $1/\lambda^2$ and therefore $\sigma \propto n_s/m^*$, provided that $\sigma(T)$ is accurately extrapolated at zero temperature. The applied magnetic field is along the c-axis and consequently, for large anisotropies, only the penetration depth parallel to the CuO planes should be involved. The experimental curves for a number of cuprates are displayed in Fig. 2.8. The critical temperature scales linearly with n_s/m^* (underdoped region), all the compounds sharing the same line; then T_c reaches its maximum value (optimally doped sample) and eventually turns back to vanishing values following a loop-shaped curve through smaller values of n_s/m^* (overdoped region). The smaller σ values correspond to an increased hole doping. The same results have been obtained

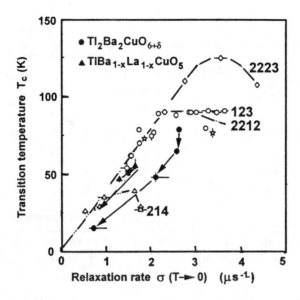

Figure 2.8: Critical temperature against the muon spin relaxation rate $\sigma \propto n_s/m_s$ for a number of compounds, n_s and m_s being the superconducting density and effective mass respectively. The arrows indicate the sequence of increasing hole doping (experiment from Ref. [Uemura91]).

by C. Niedermayer *et al.* [Nieder93]. Fig. 2.9a shows the relaxation rate as a function of temperature in the case of Tl2201 compound for different doping levels, as indicated in Fig. 2.9b, where the corresponding Hall numbers as a function of temperature are displayed.

2.4.2   Isotope effect

Isotope effect data in cuprates have been controversial for a long time. At the very beginning no correlation between critical temperature and isotope substitution had been found, leading to the conclusion that lattice excitations had not any role in the pairing mechanism. Now it is clear that T_c changes when ^{16}O is substituted with ^{18}O, while rare-earth elements substitution does not affect the critical temperature, even though recent measurements seem to support the existence of an isotope effect due to copper substitution [Crawford90,Franck91,Franck94]. BCS applied to conventional superconductors predicts $T_c \propto \omega_D \propto M^{-1/2}$, where ω_D is the Debye frequency and M is the isotope mass. Fig. 2.10 shows the oxygen isotope exponent α ($\alpha = 1/2$ in BCS case) as a function of the mobile hole concentration per CuO plane in Y123. The isotope exponent decreases as a function of doping, starting from the 1/2 BCS

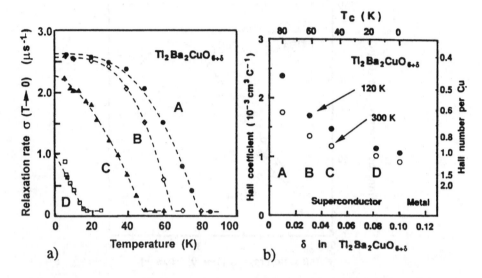

Figure 2.9: (a) Relaxation rate $\sigma \propto n_s/m^*$ vs. T for differently doped Tl2201 samples, the doping increasing from above to below [Uemura93]; (b) the corresponding Hall numbers vs. T [Kubo91].

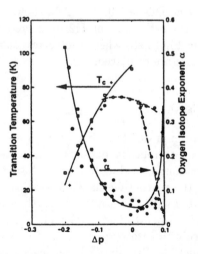

Figure 2.10: Oxygen isotope exponent α (right) for Y123 related systems vs the number of mobile holes per CuO plane, Δp. The correspondent T_c is also displayed (left). Note that $\Delta p > 0$ singles out the overdoped regime [Franck94].

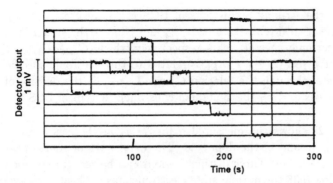

Figure 2.11: Output of a r.f. SQUID magnetometer: small integral numbers of flux quanta jumping in and out of the ring are visible. The vertical spacing corresponds to one flux quantum $hc/2e$. After Ref. [Gough87].

value at very low doping and reaching negligibly small values in correspondence of the optimally doped sample. Negative values of the oxygen isotope effect have also been reported in literature [Franck94].

2.4.3   Flux quantization

One of the striking features of the superconductors is the magnetic flux quantization. Since the system is characterized by a macroscopic wave function (the order parameter), the flux Φ of a magnetic field through the superconductor is an integer number n times a flux quantum, $\Phi_0 = hc/e^*$, h and c being the Planck constant and the velocity of light, respectively. The quantity e^* is the charge of the superconducting carriers and can be deduced from the measure of Φ. This has been done and the result for Y123 is given in Fig. 2.11, where the flux jumping in and out of a Y123 ring is shown as a function of the time. The vertical spacing is just $hc/2e$, e being the electron charge. Since the measured Φ_0 was 0.97 ± 0.04 $hc/2e$, the experiment shows that the superconducting carriers are paired electrons (or holes).

2.4.4   Specific heat

Specific heat measurements in superconductors are an important probe for electronic, phononic as well as magnetic excitations. However the high critical temperature and high critical fields H_{c2} in high-T_cSC make it difficult to disentangle the different contributions because lattice specific heat dominates at such temperatures. Furthermore, one should be able to control extrinsic contributions like those arising from defects or disorder, which can be very important in cuprates.

One of the striking predictions of BCS is on the electronic specific heat jump at the critical temperature. If C_{es} and C_{en} are the superconducting and normal electronic specific heats respectively, then $(C_{es} - C_{en})/\gamma T_c = 1.43$. The quantity $\gamma = (2/3)\pi^2 k_B^2 n_{E_F}$ is related to the electronic specific heat of a normal metal through the relation $C_e = \gamma T$ (n_{E_F} is the density of states at the Fermi level and k_B the Boltzman constant). A typical BCS value for γ is 38 mJ/mol K^2.

Fig. 2.12a shows the measured γ for $YBa_2Cu_3O_{6+x}$ as a function of the temperature T and of the doping x. The curves are of course peaked at the critical temperature. Three distinct regions can be recognized: for x approximately below 0.38 the normal state electronic specific heat varies very slightly with respect to T; for x in the 0.43-0.76 range the dependence on the temperature is more evident with a broad shoulder centered somewhere at high temperatures; finally an almost T-independent (normal metal-like) value of about 1.4 mJ/g-at.K^2 (or $1.4\times(12+x)$ mJ/mole K^2) characterizes the samples with $x > 0.8$. The three regions are evident also in Fig. 2.12b and correspond to the insulating antiferromagnetic sample, the superconducting at T_c =60K and the slightly underdoped-optimally doped sample. The same figure displays the jumps of γ at the critical temperature as a function of doping and for the fully oxygenated sample $\delta\gamma(T_c)/\gamma(T_c) \simeq 2.5$; this number is larger than the 1.43 BCS value and comparable with the 2.71, 2.37 values for Pb and Hg respectively, which are strong coupling superconductors.

As far as the low temperature specific heat is concerned, it appears to be linear in T [Overend94]. This result is contrasting with conventional materials. BCS model predicts that $C_{es} \propto (T_c/T)^{3/2}\exp(-1.76T_c/T)$, the exponential factor being due to excitations across the (isotropic) gap.

2.4.5   Superconducting gap

The superconducting energy gap Δ can be in principle measured in a number of ways. Nevertheless, in high-T_cSC the determination of Δ has been rather controversial. It is well accepted now that the ratio $2\Delta/k_B T_c$ in the ab-plane ranges from 5 to about 8, sensitively larger than the BCS value 3.52. No definite answers are yet availaible about the symmetry of the energy gap, nor on the existence of allowed states inside the gap. Both questions are very important since they are crucial points for the validity of every proposed model.

Symmetry of the order parameter

Conventional superconductors are characterized by spin singlet pairs with s-state orbital angular momentum; this condition is usually referred to as s-wave pairing. Of course other possibilities can be conceived. For instance, pairing in superfluid ^{3}He is believed to be spin triplet-driven and therefore to have

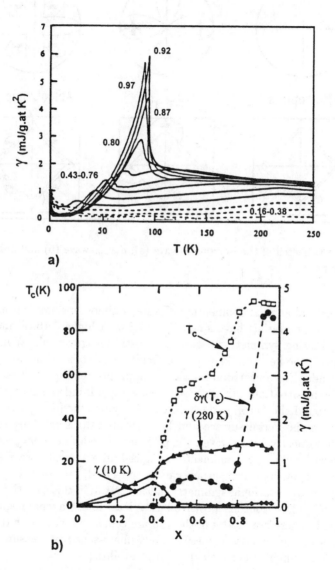

Figure 2.12: (a) $\gamma = C_e/T$ vs. T for a *single* sample of $YBa_2Cu_3O_{6+x}$ at different doping levels, x; (b) T_c, γ at T=10, 280K and jump $\delta\gamma$ of γ at T_c as a function of x [Loram93].

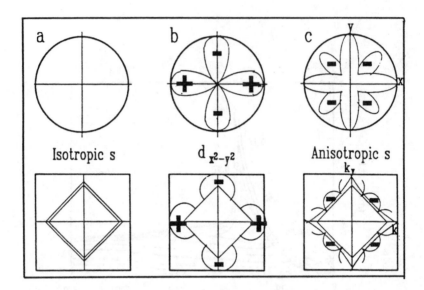

Figure 2.13: Sketch of the isotropic s-wave (a), $d_{x^2-y^2}$-wave (b) and anisotropic s (c) symmetry.

a p-wave orbital angular momentum. There is almost general consensus that heavy fermion materials have singlet pairing with higher orbital momentum (d-wave). Pairing symmetry in cuprates is still controversial. While pure s-wave seems to be very unlikely, it is difficult to establish if d-wave or the so called extended (or anisotropic) s-wave is appropriate. Fig. 2.13 sketches the three different situations. In isotropic s-wave (a) a k-independent energy gap develops around the Fermi surface; in d-wave symmetry $(d_{x^2-y^2})$ (b), the gap has nodes on the Fermi surface and lobes of opposite sign along the x- and y-direction respectively. However, the presence of nodes *does not* prove d-wave nor disprove $l = 0$ angular momentum. Indeed an extended s-wave symmetry (c) with nodes on the Fermi surface can be conceived, provided that the nodes retain the square rotational symmetry of the crystal; in this case the four lobes along x- and y-directions are positive but at least four diagonal negative lobes are required. Therefore, nodes in the gap are *consistent* with both d-wave and extended s-wave symmetry [Dynes94]. Only phase-sensitive measurements can in principle distinguish between the two possibilities.

Nodes in Δ imply the existence of electronic states inside it. This means that an arbitrarily small energy is sufficient to excite particles across the gap, in sharp contrast with isotropic s-wave superconductivity, where an activation energy exists. Along these lines, an exponential temperature dependence of a number of properties is the fingerprint of isotropic s-wave. On the contrary, a power-law dependence of the excitations is expected in case of gapless superconductivity along some directions in k-space.

Probes of the phase

The distinction between d-wave and anisotropic s-wave can be possible only through a measure of the sign of the lobes. To this aim Josephson effect appears to be suitable. Josephson [Josephson62] predicted that the tunnelling across a superconducting junction can be provided also by electron pairs. The pairs retain their identity provided that the barrier is not thicker than the coherence length. The current density J across the junction gives the difference $\Delta\phi$ in the phase of the order parameters in the two superconducting samples through the relation $J = J_c \sin \Delta\phi$.

If $d_{x^2-y^2}$-wave is involved, a π phase shift would be detected between k_x and k_y directions of the order parameter. In addition, tunnelling between a pure d-wave and an isotropic s-wave superconductor would not take place. The experiments performed so far along this direction can be of two types: the so called *weak link* experiments and Josephson tunnelling experiments. The latter are along the c-axis whereas the former probe currents in the ab-plane.

As an example of the first type, a dc SQUID (i.e. two interfering Josephson junctions) is used [Wollmann93,vanHarlingen95]. The junctions were made of Pb and high quality Y123 single crystal, the two ends facing orthogonal directions of Y123. In this arrangement, a π phase shift is expected is $d_{x^2-y^2}$-wave is involved in Y123. The same experiment has been performed using a single Josephson junction and in both cases the π shift has been detected. On the other hand the second kind of experiments supports conventional s-wave pairing. Josephson current has indeed been measured in a single junction along the c-axis (Pb-Au-high quality Y123 single crystal); this would not be the case for of d-wave symmetry. Neither this could be due to the fact that Y123 is orthorhombic rather than tetragonal: in fact the same experiment gives the same result – even though with a lower current than that expected – also in twinned samples, where the current would be averaged out [Sun94].

The word "The End" cannot yet be told about the anisotropic s- and d-wave controversy.

Tunnelling experiments

Tunneling experiments provide an useful tool to probe the existence and the magnitude of the gap in the superconducting density of states [Giaever60]. In this kind of experiments, the differential conductance of normal electrons tunnelling across the junction is measured. Different kinds of junctions can be conceived : point-contact junctions, where the tunnelling takes place between a metallic tip (MT) pushed into the superconductor (S) and the superconductor itself; planar-junctions, where a metal is evaporated on the already

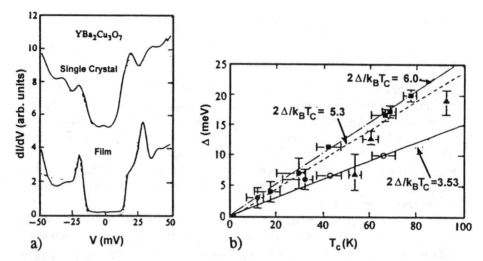

Figure 2.14: (a) Solid lines are the differential conductance from point-contact measurements on Y123 single crystal (top) and c-axis oriented film (bottom) [Kirtley90]; (b) Δ to T_c ratio for 124 and Y123 compounds; open symbols are for a planar junction with c-axis oriented films and would give Δ_c; full symbols are from point-contact tunnelling, once the contribution to Δ_c from planar junction tunnelling has been considered and therefore would give Δ_{ab} (after Kirtley [Kirtley90]).

oxydated (insulating) S-surface; S-vacuum-MT junctions; S-S junction, where the elements are two pieces of one broken superconductor sample.

Very clear experiments exist for conventional superconductors. Results on cuprates however are more difficult to be handled because of the broad transition which smears the data. In addition the tunnelling is essentially a surface probe, involving a layer of thickness $\simeq \xi$ and cuprates have a layered structure; thus it is difficult to establish if the probed layers are the CuO ones or the insulating planes. A similar difficulty is encountered in photoemission experiments. Finally, the sample anisotropy requires some modellization for the data to be reliably interpreted.

As far as the anisotropy is concerned, Fig. 2.14 shows the comparison between point-contact data for a Y123 single crystal and a c-axis oriented Y123 film. Both data refer to the differential conductance measured along the c-axis. The oriented film would give the c-axis gap.

The determination of the superconducting energy gap through this kind of measurements remains a controversial issue. In particular it appears that different families behaves in different ways: Nd- and Hg-based compounds show a well defined gap, whereas other measurements on Y123 show a gapless behaviour. This is displayed in Fig. 2.15.

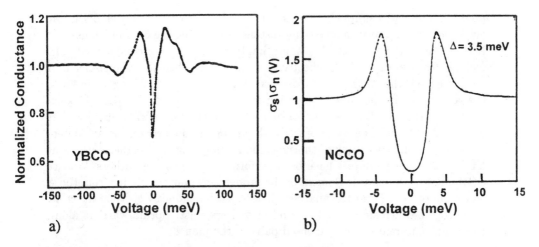

Figure 2.15: Experimental normalized differential tunneling conductance for Y123 [Valles91] (a) and $Nd_{2-x}Ce_xCuO_4$ [Tralsh91] (b).

Infrared experiments

The determination of the energy gap from infrared (IR) spectra usually hinges on the theory of Mattis and Bardeen [MattBard58] for the optical conductivity of a superconductor. Two limits are to be distinguished, the so called *clean* and *dirty* limits. In the former case $2\Delta \gg \hbar/\tau$ where τ is the Drude relaxation time ($\simeq$ the inverse of the Drude bandwidth in the optical conductivity $\sigma(\omega)$); this implies even that the mean free path is much larger than the coherence length. The opposite happens in the dirty limit. In the clean limit, below the critical temperature $\sigma(\omega)$ peaks into a delta function centered in $\omega = 0$, preventing the observation of the energy gap. In the dirty limit, the opening of the gap in the broad Drude band is visible and the missing spectral weight due to the opening of the gap is trasferred into a peak centered at $\omega = 0$; the measured reflectivity should be 1 as far as $\omega < 2\Delta$. In this case, reliable determinations of the energy gap require reflectivity uncertainties smaller than 0.5% [Calvani96] and this can be a difficult task to be accomplished. Reflectivity spectra on Y123 samples interpreted within the dirty limit picture give quite different results for the ab-plane gap to T_c ratio, ranging from 3.5 [Thomas91] to 8 [Schles90].

Many authors claim that cuprates are in the clean limit, their coherence length being quite small. Therefore any absorption edge in the optical conductivity should be due to other kind of excitations than the electronic ones across the gap; these may be lattice excitations or transitions due to localized states, or other [Calvani96].

It must be stressed that the far-infrared (FIR) reflectivity can be interpreted in terms of a simple Drude model as well. This point has been discussed very clearly by many authors (see for instance Refs. [Degiorgi89,TanTimusk92]). If the Drude model works, this would imply a gapless superconducitivity. In fact the features in the FIR reflectivity would be ascribed to the presence of a plasma edge rather than to the existence of a BCS-like gap. Along these lines, the reflectivity in the superconducting state of untwinned $YBa_2Cu_{3.5}O_{7.5}$has been interpreted by Bucher and Wachter [BuchWach92a,BuchWach92b] within a two fluid model. The poles of the real part of the dielectric constant determine the plasma frequency of the superconducting bosons. Further studies on YBCO untwinned samples have been performed by the same authors [Wachter94], allowing a disentanglement of the properties along the chains from those along the planes. From the analysis of the optical conductivity Wachter *et al.* come to the conclusion that two kind of carriers exist: normal carriers along the chains and real space condensed pairs in the planes.

Raman Spectroscopy

Raman scattering may give information on electronic excitations, since it measures the induced fluctuation of the dielectric tensor as a function of the incident light frequency. Raman spectroscopy has been used in order to obtain information on the existence and magnitude of the energy gap.

A continuous electronic scattering background is detected in high-T_c cuprates. Spectra on single crystal samples show that below a given frequency this background decreases linearly as a function of the light frequency. The extrapolated intercept at low temperature seems to be zero. In addition the background is independent of temperature in the normal state and is constant in magnitude up to large frequencies. Fig. 2.16 compares the electronic Raman scattering spectrum of Y123 with the analogous for Nb_3Sn. The latter is a superconductor known to have a well defined gap. The sharp drop in electronic scattering of Nb_3Sn below 2Δ suggests that electronic states are redistributed in the gap as far as Y123 is concerned. However a full understanding of all of the features shown by the electronic background in high-T_cSC is not available. In particular its absence in particular symmetries and its temperature independence are not clearly understood.

The experimental evidence for the gap obtained from Raman spectroscopy has been put forward by Thomsen and Cardona [ThomCard89,Thomsen90]. Phonon frequencies and linewidths in high-T_cSC are found to be anomalously temperature dependent in the vicinity of T_c . Similar changes have been observed also in conventional superconductors (Pb, for instance [Thomsen91]), the effect being 2% and 0.8% in frequency and width respectively.

In R123 a phonon softening for the 340 cm^{-1} mode is measured. The order of magnitude of the shift in frequency is 2%, depending on the rare-hearth atom (R) which substitutes Y. This large shift cannot be explained by

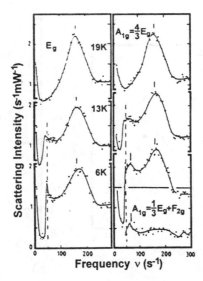

Figure 2.16: Electronic Raman scattering on Y123 single crystal [Cooper88] (left panel) and Nb_3Sn [Hackl89] (right panel) at different temperatures.

structural changes. The softening depends on the oxygen content and it is not peculiar of Raman-active modes, since also IR-active phonons with frequency ω near to 300 cm^{-1} show the same effect.

The softening is absent in non superconducting samples and reverses when the material is brought to normal state by application of a magnetic field. These circumstances suggest that the softening is related to the superconducting transition. Fig. 2.17a shows the experimental difference in the phonon frequency above and below T_c as a function of the phonon frequency itself, which is varied substituting different R atoms in R123 compound. A softening is observed until the 440 cm^{-1} phonon frequency is reached. Then a hardening happens (which is confirmed by other experiments). The solid line is a a strong electron-phonon coupling fit to the real phonon self-energy. From the data a gap position between 310 and 420 cm^{-1} can be inferred.

As far as the imaginary part is concerned, Fig. 2.17b shows the dependence of the linewidth on the temperature for superconducting Y123 as well as for the insulating parent compound $EuBa_2Cu_3O_{7-\delta}$. In the former case a broadening occurs below T_c. The solid line is a fit for the linewidth narrowing due to lattice anharmonicity effects. In the strong electron-phonon coupling model, the broadening is physically connected with the fact that a Cooper pair does not break up if the phonon energy is smaller than the pair binding energy. The onset of the broadening is therefore connected with the opening of the gap. A gap-to-T_c ratio of about 5 is then obtained.

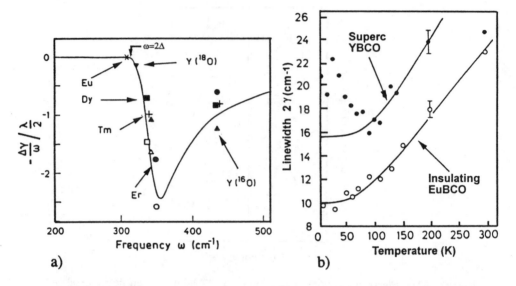

Figure 2.17: (a): Experimental phonon self-energies for different R123 compounds [Thomsen91]; open symbols are for IR-phonons. (b) Experimental linewidths vs. T for superconducting Y123 (full dots) and insulating parent compound $EuBa_2Cu_3O_{7-\delta}$ (open circles) [Friedl90].

Photoemission Spectroscopy

Photoemission Spectroscopy data give information on the position of the Fermi energy for each direction in k-space. If a gap opens, the density of states is piled up at values of the binding energies higher and lower than E_F. Handling of the data requires a careful fit. Fig. 2.18b shows the experimental data for Bi2212 below and above T_c together with the fit curve. Repeating the measure at different points in k-space gives an almost isotropic gap-to-T_c ratio of 6.8. Unfortunately this is not the end of the story. More recent measurements find a gap anisotropy, which is consistent with d-wave and anisotropic s-wave as well, as is seen in Fig. 2.18a [Shen93,ShenDessau95,Dessau93].

London penetration depth

The London penetration depth λ (see below) contains the quasiparticle density. At temperatures $T << T_c$ λ is expected to have an exponential dependence on T in the case of a conventional s-wave pairing, like in BCS model. An exponential dependence is found in Nd-based compounds, whereas a linear dependence is fitted for Y123. This result is consistent with tunnelling data, from which the Nd compound comes out to have an almost isotropic gap while Y123 seems to have nodes.

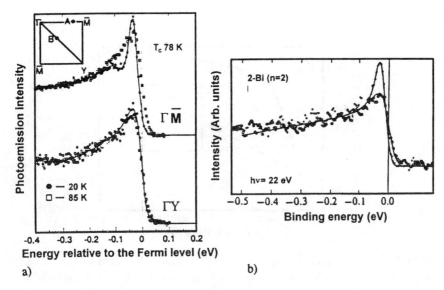

Figure 2.18: (a) High resolution ARPES for Bi2212: the gap along ΓY is absent, whereas it opens along the ΓM direction [Shen93]. Note also the dip, the origin of which is controversial. (b) ARPES data at 90K (circles) and 20K (crosses) [Olson90].

Nuclear Spin Relaxation

The ratio of the normal to superconducting nuclear spin relaxation time T_{1n}/T_{1s} in conventional BCS superconductors is expected to raise exponentially from $T = 0$ and then to peak – the so called Hebel-Slichter coherence peak – just below the critical temperature, eventually becoming 1 above T_c . This is an evidence of the existence of an energy gap, which can be fitted from the experimental data. High-T_cSC do not show any of these features for any of the nuclear sites investigated (Cu, O, etc.) (Fig. 2.19). The absence of the peak is consistent with d-wave superconductivity. However it should be noted that even in some conventional and anisotropic gap superconductors, the peak appears completely depressed. Therefore, once again, no definite conclusions can be drawn on the gap symmetry from NMR data.

2.4.6   Chemical potential

The chemical potential μ has been measured detecting the changes in the specimen work function W through a Kelvin probe method [Rietveld92]. The sample and a reference specimen (gold, for instance) make a parallel plate capacitor, the plates being connected by a wire. The measure of the current induced between the plates by the difference in work function of the two materials gives a measure of W and therefore, after calibration, of the chemical

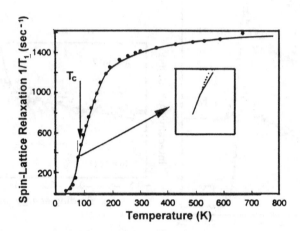

Figure 2.19: Temperature dependent Korringa ratio $1/T_1T$ data for the Cu site in $YBa_2Cu_4O_8$. Note the absence of the Hebel-Slichter coherence peak. [Machi91].

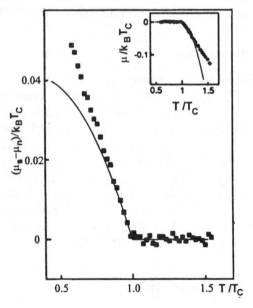

Figure 2.20: Chemical potential normalized to k_BT_c as a function of T/T_c. The solid line is a theoretical curve based on a 2D BCS model. The inset shows the same experimental data scaled with a factor -2 and a different slope subtracted in order to compare them with the theoretical curve for Bose-Einstein condensation of non interacting bosons (solid line) (experimental data from Ref. [Rietveld92]).

potential. Fig. 2.20 displays the difference between superconducting and normal state chemical potential $\Delta\mu = \mu_n - \mu_s$ normalized to $k_B T_c$ as a function of the normalized temperature T/T_c. The main body of the figure and the inset show the data compared with a 2D BCS local-pairing model and the Bose-Einstein condensation for an ideal gas, respectively. The main conclusion is that the chemical potential has a jump in slope at the critical temperature $d\Delta\mu/dT|_{T_c} = 0.12{\pm}0.02$. It is noteworthy that it can be shown on very general grounds [Marel92] that the jump in slope of $\Delta\mu$ is directely connected to the jump in the specific heat through the variation of T_c upon carrier density changes:

$$\frac{d(\mu_n - \mu_s)/dT}{C_n - C_s}\Big|_{T_c} = \frac{d\ln T_c}{dn}.$$

2.4.7   Magnetic properties

Superconductivity can be destroyed applying a magnetic field larger than a critical value, which is a decreasing function of the temperature T, vaninshing at and above T_c.

High-T_cSC are all type II superconductors. This means that actually two critical magnetic fields exist, usually referred to as H_{c1} and H_{c2}, with $H_{c1} < H_{c2}$. The magnetic field is completely excluded from the superconductor only if it is smaller than H_{c1}. Fields with values between H_{c1} and H_{c2} penetrates in vortices, the fluxoid Φ_0 being the quantum associated to each vortex. The number of vortices increases as far as the magnetic field approaches H_{c2} and then it penetrates completely the material.

Lower critical field

The lower critical field at $T = 0$ can be of the order of 0.8 Tesla along the c direction and a factor 4 smaller in the ab plane, this anisotropy remaining constant as a function of the temperature. Fig. 2.21a displays the described behaviour in the case $YBa_2Cu_3O_{7-\delta}$.

London penetration depth

Magnetic field is excluded from the superconductor bulk exponentially decreasing away from the surface. The length scale λ for such a decay is called London penetration depth and it depends on temperature, diverging at T_c. In pure superconducting metals, $\lambda(T = 0)$ is of the order of 500 $\overset{\circ}{A}$, while in high-T_cSC it can be as high as 1400 $\overset{\circ}{A}$ in the ab plane.

Within a two fluid model, in which superconducting as well as normal carriers exist at finite temperature, the expected temperature dependence of λ on T is $\lambda(T) = \lambda(0)[1 - (T/T_c)^4]^{-1/2}$. $\lambda(0)$ is defined by $\lambda^2(0) = mc^2/(4\pi n_s q^2)$, m, n_s and q being respectively the mass, the density and the charge of the

Maria Luisa Chiofalo

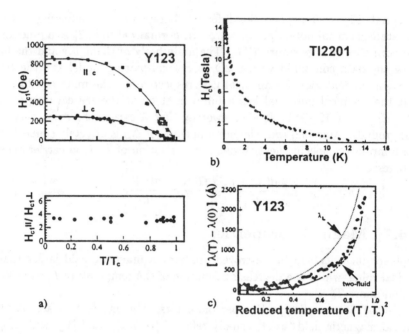

Figure 2.21: (a) Lower critical field along c and ab directions (left) and their ratio vs. T (right) for Y123 [Wu90]; (b) c-axis upper critical field as a function of T for T2201 [Mackenzie93,Osofsky93]; (c) London penetration depth vs. T for Y123 [Krusin89].

superconducting carriers and c being the velocity of light. Actually the previous formula does not tell all the story; if ξ is the coherence length and l the mean free path, a factor $(1 + \xi/l)$ should multiply the prevoius equation for λ. However, all high-T_cSC are estimated to have $\xi/l < 0.3$, namely they are claimed to be in the clean limit. Fig. 2.21c shows the measured λ for Y123 as a function of the temperature and its comparison with the two fluid model prediction.

As a final remark, in an isotropic superconductor H_{c1} and λ are closely related because $H_{c1} \simeq \Phi_0/4\pi\lambda^2$. In the anisotropic case the relationship requires some care.

Upper critical field

H_{c2} at $T = 0$ in high–T_cSC assumes really huge values up to several tens of Tesla. This fact has made difficult for a long time a reliable determination of the temperature dependence of the upper critical field far from T_c . Recently it has been possible to measure such a dependence in overdoped Tl2201 and in Bi2201 [Mackenzie93] samples. In both cases the critical temperature is not very high and the upper critical fields can be produced without major problems; furthermore the field transition is not as broad as it usually is. Fig. 2.21b shows the experimental result concerning Bi2201 together with the theoretical predicted behaviour. The striking result is the upward curvature of H_{c2} .

Coherence length

High-T_cSC are characterized by a very short coherence length, as compared to conventional superconductors. This implies a Ginzburg-Landau parameter $\kappa = \lambda/\xi$ much larger than the Abrikosov critical value $1/\sqrt{2}$; therefore, as already noted, high-T_cSC are extreme type II superconductors.

The coherence length ξ in cuprates is derived from the measure of the upper critical field through the relation $H_{c2}(T = 0) \simeq \Phi_0/2\pi\xi^2(0)$, where ξ is in the plane perpendicular to the direction of the magnetic field. The previous relationship is valid, in that form, only in the isotropic case. From H_{c2} measurements it turns out that typical values of ξ in the ab plane are within the 10-30 Å range. For instance, ξ_{ab} in Y123 is 16 Å, whereas $\xi_c \simeq 3$Å, about a factor three smaller than the c-axis length, i.e. the distance between two sets of immediately adjacent CuO planes. It has to be noted that, as far as the critical temperature is approached, the coherence length becomes larger and diverges at the critical temperature in such a way that the pairs cannot longer be considered as confined in the planes.

In the anisotropic case, the relationships among the lower and upper critical fields and the London penetration depth and coherence length in the different directions are a bit more complicated and all the quantities are mixed together.

2.5   Normal-state properties

2.5.1   Transport properties

Resistivity

The resistivity in cuprates has a peculiar dependence on temperature. Fig. 2.22a shows that the ab-plane resistivity is approximately linear over the whole range of temperatures for the members of the family with the highest transition temperatures. This result has to be contrasted with the usual normal metal resistivity, which is given by the Bloch-Grüneisen formula $\rho \propto T^5$ at temperatures $T << 0.2\Theta_D$, where Θ_D is the Debye temperature, and is linear only for $T > 0.2\Theta_D$. Departures from the linear behaviour are observed for instance in the case of Nd- and La- based compounds. Moreover it is worthwhile to mention that the in-plane resistivity of $YBa_2Cu_4O_8$ is fittable with the Bloch-Grüneisen formula [Berghuis90]. Anyways, the linear dependence is mainly

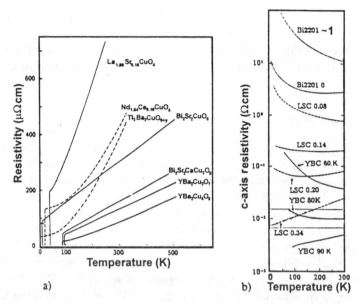

Figure 2.22: (a) Temperature dependence of the in-plane resistivity for a number of high-T_c cuprates [Iye92]; the absolute values are semi-quantitative. (b) ρ_c for different compounds (after Ref. [Ito91a,Ito91b]).

peculiar of the (nearly) optimum doped samples. In the so called *overdoped* region, i. e. where T_c falls down from the maximum value at optimum doping, the dependence is still power law-like, $\rho_{ab} \propto T^\gamma$; however in this case the exponent γ varies from 1 to 2 as far as the doping is increased [Iye92]. The T^2 dependence would be consistent with a dominant contribution from electron-electron scattering within a Fermi-liquid picture.

An exception to the cited behaviour seems to be the fully oxygenated Y123 sample. However Y123 compound is peculiar in the cuprates panorama, because it has the chains which are fully occupied in the fully oxygenated (YBCO$_7$) case; keeping in mind that the b-axis conductivity has contributions from the plane and from the chains whereas the a-axis one contribute only to the planes, a straightforward analysis of the data shows that about the 60% of the conductivity comes from the chains [Friedman90]. A direct evidence of this fact comes from the experiments of Bucher and Wachter [BuchWach95] who have investigated the transport and optical properties of YBa$_2$Cu$_{3.5}$O$_{7.5}$. In this case the contribution from the a and b-axis can be more easily disentangled because YBa$_2$Cu$_{3.5}$O$_{7.5}$ is untwinned. Thus the properties measured along the a-direction can be considered as due only to the planes. The resistivity along the chains shows a T^2 dependence on the temperature up to 300 K. The resistivity in the planes has an unconventional behaviour, it being linear in T for temperatures larger than about 160 K.

As far as the c-axis resistivity is concerned, it often shows an upward curvature near T_c which is reminiscent of a semiconductor-like behaviour (Fig. 2.22-(b)). However, if this would be actually the case, the temperature dependence would be exponential, like for a thermally activated process; on the contrary, ρ_c still has a power law dependence with γ varying from 0 to 2 depending on doping [Carrington93]. The upward curvature tends to disappear with increasing the doping, when the anisotropy reduces too.

Two further comments would be interesting. First, the angular coefficient in the linear term is almost the same for all of the compounds and is of the order of $0.5\mu\Omega$-cm, indicating a common scattering mechanism. Secondly, the values of the in-plane resistivities are as high as some tenth of mΩ-cm and the out-of-plane ρ are of the order of some Ω-cm; this leads to a factor 100 in the anisotropy ρ_c/ρ_{ab}; as already noted, the anisotropy reduces in the overdoped regime. In Y123 an additional anisotropy in the ab plane $\rho_a/\rho_b \simeq 2.2$ is due to the presence of the chains.

Summarizing, the resistivity of cuprates has a power law dependence on temperature, the exponent ranging from 1 to 2 for ρ_{ab} and from 0 to 2 for ρ_c. The exponent varies upon overdoping the sample, the optimum doping ρ_{ab} being linear in T. It is not at all clear if the unusual behaviour of the resistivity in cuprates is built up with their anisotropy and nearly two– dimensional character or it comes about from a particular microscopic mechanism. In particular no correlation appears between T_c and the c-axis resistivity; thus any microscopical mechanism which is based on interplanar coupling seems to be disfavoured. Finally, any three-dimensional phonon scattering contribution to the resistivity appears to be irrelevant as far as the critical temperature is smaller than the Debye temperature. On the other hand it is not clear how meaningful is the Debye approximation in materials so rich in optical phonons.

2.5.2   Existence of a Fermi surface

Band structure calculations

Fig. 2.23 displays the first Brillouin zone of the relevant structures found in cuprates. The calculated one-electron band structure of Bi2212 together with its Fermi surface is shown in Fig. 2.24. The bands along the ΓZ direction (right part of the figure) are flat, as expected since this corresponds to the out-of-plane direction. As far as the ab-plane is concerned, the Fermi energy (dashed line) crosses the bands in ΓX and ΓM directions. The bands are due to the hybridized CuO orbitals. Pockets arise in the middle of the ΓZ direction and are due to BiO hybridized states. They are absent in ARPES experiments (see below).

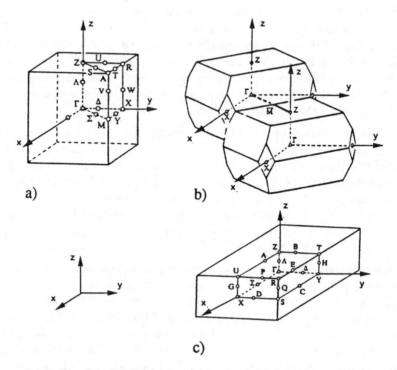

Figure 2.23: The first BZ for (a) pt, (b) bct and (c) primitive-orthorhombic lattices [Burns92].

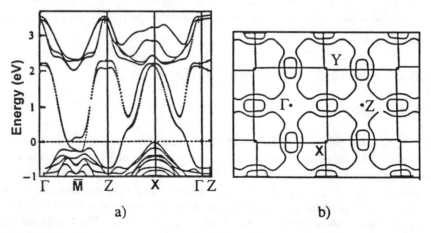

Figure 2.24: Band structure calculation (a) and Fermi surface (b) for Bi2212 [Pickett89].

Photoemission Spectroscopy

The single-particle excitation spectrum can be probed by means of photoemission spectroscopy. Photons of appropriate energy $\hbar\omega$, depending on the electronic levels to be probed, impinge the surface sample and electrons are emitted with a kinetic energy $E_{kin} = \hbar\omega - E_b - w$, where E_b is the electron energy measured with respect to the chemical potential and w is the work function of the crystal. The emitted electrons, before escaping from the surface, may suffer collisions with other electrons. Therefore, even though light penetrates deeply into the sample, the electrons which are emitted come from a very thin shell below the surface, typically 0.5-2 nm. Very clean surfaces and ultra-high vacuum experimental conditions are needed.

Once the work function has been measured, it is possible to determine the density of states of valence electrons or the core level electron energies if the electrons at all angles are detected; this corresponds to an angle-integrated photoemission. In Angle-Resolved Photoemission Spectroscopy (ARPES) the electron band dispersion energy $E(\mathbf{k}_{\|})$ can be measured as a function of the electron k-vector parallel to the surface, which is the conserved component in the process.

The most studied cuprate is Bi2212 since it can be grown in large crystals. However some aspects are shared by the whole cuprate family. These aspects can be summurized as follows: a Fermi surface exists (Fig. 2.25a); flat bands are present close to the Fermi energy E_F for hole doped and far from E_F in electron-doped cuprates (Fig. 2.25b); the lineshapes near E_F seem to be unusual, the inverse quasi-particle lifetime turning out to be $\propto \omega$ rather than ω^2 like in a Fermi liquid. The last result must not be considered conclusive [Shen93]. All the cited results but the last one are in general agreement with one-electron band structure calculations. However it has to be noted the fact that the ARPES band mass along the ΓY direction is a factor 2 to 5 larger than the calculated one. Furthermore doping effects appear to be important in ARPES; the band structure change on the overall energy range and a simple rigid band shift cannot account for the modifications [Takahashi89].

2.5.3   Electron-phonon interaction

The relevance of electron-phonon effects for high-T_c superconductiviy is remarked by a number of experimental results. Among them the isotope effect (see sec.2.4.2); the existence of large splittings of the longitudinal-transverse optical phonon and by polaronic features in the optical conductivity, surviving also in the metallic state (sec.2.5.3); the polaronic nature of photoinduced carriers in parent non-superconducting compounds (sec.2.5.3); the data from tunnelling experiments, where structures in the imaginary part of the gap function appear in correspondence of the relevant phonon energies (sec.2.5.3);

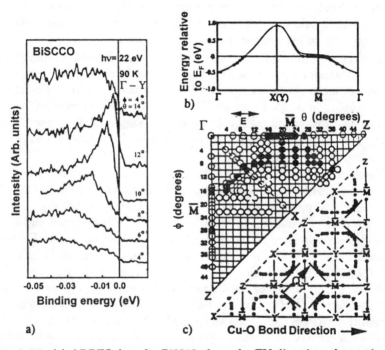

Figure 2.25: (a) ARPES data for Bi2212 along the ΓY direction; the vertical line indicates the position of the Fermi energy [Olson90]. (b) Experimental ARPES band structure for Bi2212 [Shen93]. (c) ARPES Fermi surface for Bi2212 [Dessau93]; note the absence of the pockets with respect to the band structure calculation in Fig. 2.24.

the enhanced effective carrier mass resulting from London penetration depth measurements (sec.2.4.7,6.4).

Phonon density of states

Cuprates have many atoms in the unit cell and therefore their phonon structure is rather rich. As an example, the generalized phonon density of states for Y123 is shown in Fig. 2.26 together with the indication of Raman and infrared spectroscopy findings. Generally speaking, the modes with frequencies smaller than about 300 cm^{-1} involve the motion of the rare earth atoms whereas the higher modes involve the oxygen atoms in the copper planes and in the apical position as well.

Some general remarks are to be made. First, because of the high conductivity in the planes, the light penetrates a very short depth and therefore phonons polarized along the ab-plane are hardly seen in the metallic phase by optical spectroscopy. Additional information in the planes can be obtained whenever a lower density parent compound is available. As an example we remind the case of Pr123, which has the same lattice structure as Y123 but lower ab-plane conductivity; therefore the transverse optical phonons are very similar in the two compounds and are visible even by optical spectroscopy. However the longitudinal optical phonons are in general rather different because of the coupling with plasmon modes.

Secondly, Inelastic Neutron Scattering (INS) results are strongly dependent on the availibility of large samples and only recently some reliable data have been obtained. They are not discussed in this section, but three points can be mentioned [PintReich94]: (i) the measured spectra are in good agreement with phonon calculations; (ii) there is a general agreement with optical spectroscopy and specific heat measurements but, (iii) at variance with the other techniques, no appreciable changes in the lattice vibrations at T_c are detected [Renker88]. The latter puzzling result is evident in Fig. 2.26a, where nor shifts in phonon frequencies neither changes in linewidths appear upon cooling the sample. On the other hand changes in the phonon density of states have been measured between the superconducting Y123 and the semi-insulating $Y_{1-x}Pr_xBa_2Cu_3O_{7-\delta}$ [Renker88]; this is shown in Fig. 2.26b, where the difference between the two PDOS has strong enhancements at about 20 and 50 meV, which are the frequencies of O_p-driven phonons. If one believes that the substitution of Y with Pr does not affect the phonons due to the motion of the oxygens in the CuO planes, the previous result is a strong indication of electron-phonon coupling.

LO-TO phonon splittings and ionicity

Large longitudinal-transverse (LO-TO) splittings are found for optical phonons [Noh88]; this is an indication of the strong ionicity of the material.

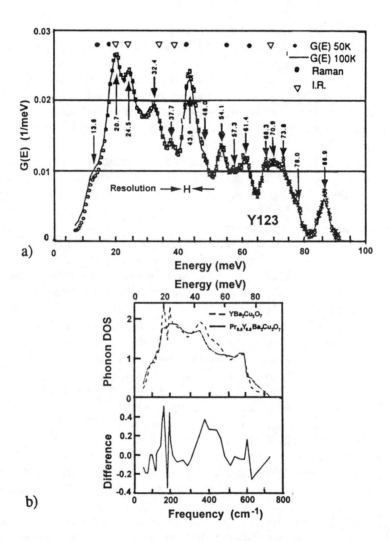

Figure 2.26: (a) Generalized Phonon Density of States (GPDOS) for Y123 at optimum doping at 50 and 100K [Arai92]. Note that on the generalized PDOS the contribution of the different species is differently weighted and a reliable lattice dynamical model is required to extract the PDOS. (b) GPDOS of $YBa_2Cu_3O_7$ and $Y_{1-x}Pr_xBa_2Cu_3O_{7-\delta}$ (upper panel) and their difference (lower panel) [Renker88].

Mode	KK analysis		Osc. fit			Theory		
	ω_{TO}	ω_{LO}	ω_{TO}	γ_{TO}	S	ω_{TO}	ω_{LO}	S
A_{2u}	108	108	105	5	2.2	116	118	2.0
A_{2u}	156	187	145	12	3.0	153	163	3.3
A_{2u}	216	255	210	6	0.5	189	203	1.6
A_{2u}	370	466	365	20	2.1	390	455	1.7
A_{2u}	640	661	650	20	0.15	540	543	0.3
E_u	65	72				55	62	2.5
E_u	120	122	115	4	2.0	105	106	0.1
E_u	189	198	188	6	1.8	214	243	2.1
E_u	248	265	244	5	1.2	276	318	0.7
E_u	355	416	350	18	2.2	355	414	1.5
E_u	579	629	596	17	0.5	502	507	0.6

Table 2.3: Experimental and calculated tranverse (ω_{TO}) and longitudinal (ω_{LO}) optical frequencies, damping (γ_{TO}) and oscillator strengths (S) of IR active phonons in $YBa_2CU_3O_6$ [Cardona94]. Frequencies and dampings are in cm^{-1}. The theoretical quantities as well as the experimental results from oscillator fits are after Ref. [ThomCard89]; experimental data from Kramers-Kronig analysis are after Ref. [Bazhenov89].

For instance, Tab. 2.3 lists the c-polarized phonon frequencies of insulating $YBa_2Cu_3O_6$ together with their respective oscillator strengths S_i. The oscillator strength is connected to the LO-TO splitting of the optical phonon and the larger S, the larger is the splitting [Cardona94]. In terms of macroscopical quantities, the ratio of the high frequency to the static dielectric constant is $\epsilon_\infty/\epsilon_0 = \Pi_i^n(\omega_T^i/\omega_L^i)^2$ [AshMerm], $\omega_{T,L}^i$ being the transverse and longitudinal frequencies of the oscillator i. From the data reported in the table, $\epsilon_\infty/\epsilon_0$ is of the order of 0.1. Appreciable splittings survive also in the (doped) $YBa_2Cu_3O_7$ [Cardona94], even for ab-plane polarized phonons.

Assignment of the modes

The assignement of the phonons comes through neutron scattering (NS) experimental data and through the comparison with lattice dynamical calculations. As an example, Fig. 2.27 shows the infrared (IR) (ungerade subscript) modes of Y123 together with the experimental IR and INS data.

The higher c-polarized modes (the last two in the first row in Fig. 2.27) involve respectively the motion of O_p oxygens (oxygens in CuO planes) and the motion of apex oxygens. These two oxygen atoms are responsible also for the strongest ab-polarized modes (the fourth and the fifth from the left in the

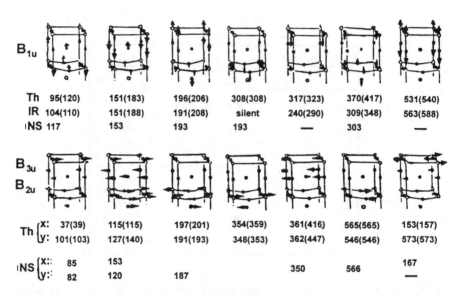

Th	95(120)	151(183)	196(206)	308(308)	317(323)	370(417)	531(540)
IR	104(110)	151(188)	191(208)	silent	240(290)	309(348)	563(588)
INS	117	153	193	193	—	303	—

Th { **x:** 37(39)   **y:** 101(103) | 115(115)   127(140) | 197(201)   191(193) | 354(359)   348(353) | 361(416)   362(447) | 565(565)   546(546) | 153(157)   573(573) }

INS { **x::** 85   **y::** 82 | 153   120 | 187 | | 350 | 566 | 167   — }

Figure 2.27: Calculated ungerade eigenmodes of Y123 (O_7) together with experimental IR and INS data [Cardona94]. Frequencies are in cm^{-1}. For atom identification, see Fig. 2.5.

last row of Fig. 2.27); the two modes are at about 270 and 340 cm^{-1} and have the largest LO-TO splittings.

Infrared Spectroscopy

One and two component picture of the normal-state ab-plane conductivity. infrared spectroscopy has been widely used in the investigation of high-T_cSC . As far as the normal state optical properties are concerned, one of the most remarkable characteristics of the in-plane optical conductivity $\sigma(\omega)$ is the deviation from the $1/\omega^2$ Drude law, typical of Fermi liquid behaviour for frequencies ω much larger than the inverse relaxation time $1/\tau$. This is clearly visible in Fig. 2.28 for some cuprates.

Two approaches have been used to explain such an anomalous behaviour [Calvani96]. Remaining within a standard Drude-Lorentz model, the experimental data have been interpreted in terms of a multicomponent picture. This is shown in Fig. 2.29: the whole conductivity for Bi2212 displayed in Fig. 2.29a at different temperatures is fitted using three different contributions. One of them is the Drude conductivity, with $1/\tau \propto T$, keeping τ and the optical mass independent of frequency (Fig. 2.29b); the Drude term peaks and narrows on cooling, condensing into a $\delta(\omega = 0)$-like function. The other two terms are visible in Fig. 2.29c, after the Drude peak has been subtracted. This broad feature has an onset at 150 cm^{-1} and is peaked at about 1000 cm^{-1}.

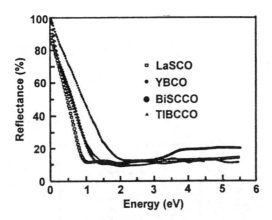

Figure 2.28: Infrared and ultraviolet reflectance of 124 (LaSCO), Y123 (YBCO), Bi2212 (BiSCCO) and Tl2223 (TlBCCO) systems [TanTimusk92]. The linear behaviour vs ω is evident.

It is composed by the temperature dependent peaked part, usually referred to as d-band, while the second term is given by a temperature independent tail, usually called Mid-Infrared Band (MIR). The previous description can be considered to be the typical of the optical conductivity for cuprate materials.

A different modellization for the anomalous frequency dependence of the optical conductivity lies in the so called anomalous Drude model, in which the relaxation time as well as the carrier mass in the Drude model are taken to be frequency dependent. It has to be noted however that the fits to the anomalous Drude model have some problems at lower temperatures [Calvani96].

The d-band in optical conductivity and polaron effects. Focusing on the normal Drude interpretation of the optical conductivity, the origin of the d-band has to be addressed. Studies on metallic and semiconducting parent compounds suggest that the d-band is due to polarons. The evidence of such an origin stems from the analysis of the optical conductivity in the insulating as well as in the metallic phase. Fig. 2.30ab shows the reflectivity of $BI_2Sr_2YCu_2O_8$ and Gd_2CuO_4, which are insulating parent compounds of Y123 and Nd_2CuO_{4-y} respectively. The d-band is shown at different temperatures and has additional superimposed infrared modes (IRAV); the latter differ in energy by about the characteristic phonon frequency of the material and follow very closely the underlying d-band as far as the temperature and doping dependence are concerned [Calvani96]. The additional IR-active modes survive even in the metallic phase, as shown in Fig. 2.30c in case of Bi-based metallic compounds. Similar results have been obtained also for Y- and Nd-based materials and for 124 compound [TanTimusk92,Uchida91].

Photoinduced absorption on insulating parent compounds. An important class

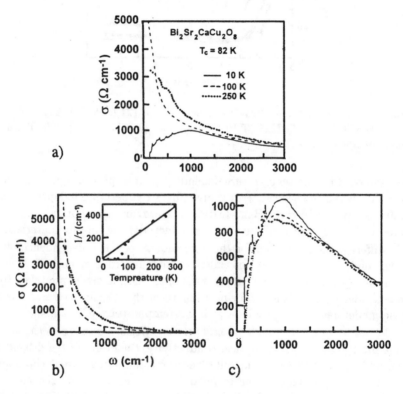

Figure 2.29: (a) Optical conductivity for Bi2212 single crystal at T=10, 100, 250K; (b) the fit to the Drude part of (a) and, in the inset, the fitted temperature dependence of $1/\tau$; (c) the remaining MIR and d-(or J-)bands after subtraction of (b) from (a) [Thomas91].

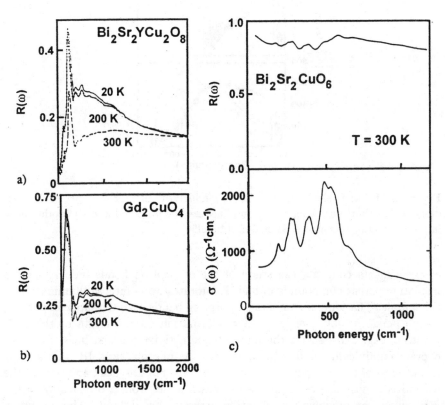

Figure 2.30: Reflectivity of (a) $Bi_2Sr_2YCu_2O_8$ and (b) Gd_2CuO_4 [Calvani] at different temperatures. (c) Reflectivity (top) and conductivity (bottom) of metallic $Bi_2Sr_2CuO_6$ thin film, after Ref. [Calvani96].

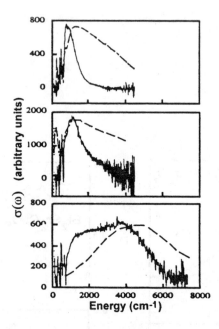

Figure 2.31: (a) Dashed lines are the IR conductivity of Tl2212 (top), Y123 (middle) and 124 (bottom), whereas the full lines are the photoinduced IR conductivity in their insulator precursors, after Ref. [Mihail90].

of experiments concerns the study of the absorption bands induced in the insulating parent compounds of high-T_c cuprates upon intense illumination of the sample. The comparison of the absorption bands with the superconducting sample analogs may give important information on the origin of the MIR features. Fig. 2.31a shows the infrared conductivity (dashed lines) for three cuprate family compounds compared with the photoinduced IR conductivity (solid lines) of their respective insulator precursors. The peak and part of the d-band is almost unaltered in the two cases–apart from a broadening in the MIR side. The previous consideration suggests that d-band is due to lattice distorsions induced by the carriers upon doping.

The plasmon edge and the carrier mass. Fig. 2.32 displays the real part of the dielectric function for a 124 sample at different doping levels. The frequency at which the real part of the dielectric constant is zero corresponds to the screened plasma frequency ($\omega_p/\sqrt{\epsilon_\infty}$). As it can be seen, the screened plasma frequency increases with increasing doping up to $x \simeq 10$ and then remains almost constant. This means that the carrier density n divided by the effective mass m^* does not change appreciably upon further doping. If the analysis is correct, this implies a density dependent effective mass which decreases less than n and almost like n for $x < 0.1$ and $x > 0.1$ respectively.

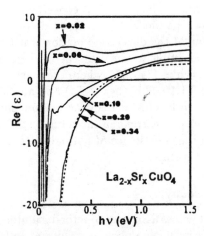

Figure 2.32: Real part of the dielectric function as fitted from reflectivity data on a 124 compound at different doping x [Uchida91].

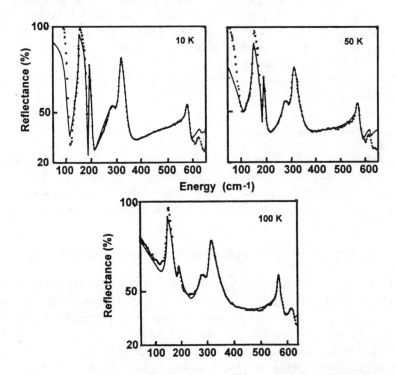

Figure 2.33: Reflectivity data of Y123 in E∥c configuration and at temperatures T=10, 60, 100K. Points are the experimental data and solid lines the fits to a model dielectric function which contains LO phonon-plasmon coupling. After Ref. [Gajic92].

Plasmon-phonon coupling. A further result concerns the c-axis polarized spectra. It has been shown [Gajic92] that a good fit of c-axis reflectivity data requires a modellization of the dielectric function which includes LO phonon-plasmon coupling in addition to the free carrier background and the MIR component. This is shown in Fig. 2.33, where the plasma edge is visible at 365 cm^{-1} at temperatures T=10, 60K (below T_c) and at 600 cm^{-1} at $T = 100$ K (above T_c). The change in plasma frequency below and above T_c tells that about the 60% of the charge carriers in the normal state are condensed below T_c ; this fact suggests the coexistence of normal state and condensed carriers below the critical temperature.

Tunnelling Spectroscopy

The tunnelling current across a Superconductor-Insulator-Superconductor (SIS) junction may give information on phonon structure. In fact if some excitation interacts with the superconducting quasi-particles, the latter can be strongly damped and an imaginary part of the energy gap $\Delta(E)$ develops. Thus, peaks in Im(Δ) correspond to the energy of the excitations involved. In conventional Pb-I-Pb junctions such an effect has been seen at energies of Pb phonons, thus demonstrating that phonons play a role in mediating the pairing. Similar results have been found in Bi2212-I-Bi2212 junctions, the peaks being in agreement with the phonon spectra in Bi2212 [Miyakawa89], yielding a gap-to-T_c ratio of about 6.

Concluding remarks

Concluding this part, IR and Raman spectroscopy as well as tunnelling experiments reveal that electron-phonon coupling is important in cuprates. The comparison of the spectra with the insulating related compounds suggests that electron-phonon interaction plays a role in the superconducting pairing.

2.6   A reference review of theoretical models

Many theories have been proposed, among which the Resonating Valence Bond theory by Anderson [Anderson87a,Anderson87b], purely electronic mechanisms like the exciton coupling one by Varma *et al.* [Varma87], the spin fluctuation model by Pines and coworkers [Pines90], spin bags by Schrieffer *et al.* [Schrieffer88], anyon superconductivity [Wilczek91] and bipolaron models, first proposed by Alexandrov and Ranninger [AlexRann81a,AlexRann81b,AlexRann86]. On a less "exotic" side, some attempts have been made in order to go beyond the Eliashberg theory [Eliash60] for strong electron-phonon coupling and to overcome the Migdal theorem limitations [Migdal58,PietroStrass92]. Remaining within the Eliashberg formalism, it has been shown that the right T_c values and other doping dependent properties can be obtained, provided that E_F pins on a logarithmic van Hove singularity in the density of states, as it appears from band structure calculations [Tsuei90]. In this case there is no need for unusually large electron-phonon coupling parameters.

The problem of polaron and bipolaron formation has been faced by many authors [AlexMott93,Mott93,AlexMott94a], [DeJongh88,Micnas90], [Emin89,- EminHill89], [Bassani91,Devreese91]. From a general point of view, bipolaron superconductivity hinges on Bose-Einstein Condensation of charged bosons [Scha55,Scha57], which had already been studied in connection with conventional superconductivity. Along these lines, it has been shown that the condensation of charged bosons is consistent with many general properties of superconductors [Scha55,AlexRann81,Chak87].

A further important research field is given by the phenomenological theories of crossover from BCS to Bose-Einstein condensation (BEC). For instance it has been shown that both BCS and BEC can be recovered changing the dilution parameter of the system ($nR_p^3 \ll 1$ and $nR_p^3 \gg 1$ respectively, where n is the electronic density and R_p the average radius of the pair) [NozSchmit-t85,Nasu87,Pist93]. Moreover it was also shown that, in two- dimensions, the existence of a bound state makes the many particle system unstable versus a superconductivity state; furthermore, if the binding energy is large with respect to the Fermi energy, one has a Bose condensation and in the opposite limit a BCS state [Randeria90]. Finally, we want to remind the boson-fermion models by Ranninger and Robaszkiewicz [RannRobasz85] and by Friedberg and T.D. Lee [FriedLee89a,FriedLee89b], which will be considered in some detail in chapter 7.

In the next section we shall focus on bipolaronic superconductivity, where the so called small and large polaron approaches have been considered.

The concept of polaron was first introduced by Landau [Landau53] in relation to electrons in the conduction band of ionic crystals. Due to the ionic effects and to the presence of core electrons, the effective dielectric constant ϵ^* of the medium can be written as

$$\frac{1}{\epsilon^*} = \frac{1}{\epsilon_\infty} - \frac{1}{\epsilon_0} .$$

The potential energy of the electron of charge e as a function of the distance r is therefore given by $-e^2/(\epsilon^* r)$. For an ionic lattice, the gain in energy of the electron confined within a polarization region of size R_p (the polaron radius) is $E_p = e^2/(\epsilon^* R_p)$. A polaron is adiabatic when the electron bandwidth is larger than the typical energy of the phonon involved in the electron-lattice interaction. The total energy of an adiabatic polaron has two types of minima with respect to the size of the polarization cloud; they depend on whether the polaron radius is smaller or larger than the lattice constant, yielding a description in terms of small or large polarons in either cases [Emin93]. The main concepts related to this subject are reviewed below.

2.6.1   Small (bi-)polarons

Small polaron theory was first considered by Anderson [Anderson75] and Street
and Mott [StreetMott75]. Small bipolaronic superconductivity has been considered by Alexandrov, Ranninger [AlexRann81], Micnas [Micnas90], Mott [Mo-
tt93]. A small polaron is an electron strongly bound to a specific lattice site
by a strong short range electron-lattice interaction. Therefore a small polaron
can form when the self-energy correction due to the electron-lattice interaction
is larger than the hopping integral J between nearest neighbouring sites in a
tight binding picture [AlexMott94b,AlexKrebs92, Emin89,Mahan,Devreese96].
Correspondingly, the bare bandwith renormalizes and narrows exponentially
due to lattice effects. Furthermore, the values that the average polaron radius
can assume is bounded from above, as is discussed later on. Given two of these
small polarons on adjacent sites in the tight binding picture, a small bipolaron
forms when the attractive interaction due to the quasistatic lattice deforma-
tion overwhelms the Coulomb repulsion V_c; if E_p is the polaron self-energy,
this happens whenever $2E_p - V_c > 0$, which has to be considered the small
bipolaron binding energy. Alexandrov and Ranninger [AlexRann81] have introduced the concept of mobile small polarons and bipolarons. Small polarons
and bipolarons can move coherently though with a huge effective mass [Ku-
perWhit63] at temperatures smaller than half the typical phonon energy ω_0,
namely $k_B T < \hbar \omega_0 / 2$. At larger temperatures the motion is due to incoherent
activated hopping from site to site.

   Some parameters can be introduced in order to fix the previous concepts [Al-
exMott94b]. If D is the bare electron bandwidth and z the number of nearest
neighbours ($z = 6$ for a cubic lattice), then the hopping integral is $J = D/2z$.
The condition for small polaron formation then reads $E_p/D \gg 1/2z$. The
quantity $\lambda = 2z E_p/D$ is the expansion parameter of the theory; it is practically
equivalent to the strong coupling parameter used in superconductivity, namely
$\lambda = N(0)V_0$, where $N(0)$ is the density of states at the Fermi level and V_0 the
typical strength of electron-phonon interaction. The expansion parameter λ
is connected to the adiabaticity ratio $\omega_0/E_F \simeq \omega_0/D$ of the typical phonon
energy to the Fermi energy. In particular the polaronic shift E_p is given by
$E_p = g^2 \omega_0$, g^2 being an adimensional electron-phonon coupling constant which
measures the average number of phonons in the polaron cloud; therefore λ is g^2
times the adiabaticity ratio. Among the parameters g^2, λ and ω_0/E_F, clearly
only two of them are independent and $\lambda = g^2 \omega_0/E_F$ [AlexMott94b].

   First it is important to note that the condition ω_0/E_F larger or of the order
of 1 states the violation of Migdal's theorem, which the BCS strong coupling
theory of superconductivity hinges on. Let us call λ_c the critical λ for small
polaron formation. Therefore the condition for small polaron formation is $g^2 >
\lambda_c D/\omega_0$. The calculated values of λ_c as a function of ω_0/D [AlexKrebs92] are
shown in Fig.2.34, together with the correspondent polaron mass enhancement;

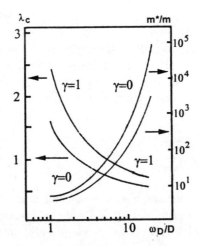

Figure 2.34: Critical values of the effective coupling constant λ_c and correspondent mass renormalization m^*/m as a function of the ratio ω_D/D (see text; calculation after [AlexKrebs92]). $\gamma = 0$ corresponds to a local interaction whereas $\gamma = 1$ to a long range Fröhlich interaction.

the calculation is supported by Monte Carlo simulation results by de Raedt and Lagendijk [RaedtLagen84]. Small polarons may form even when Migdal theorem applies, i. e. $D/\omega_0 > 1$, provided that g^2 is sufficiently large. As an example, for $\omega_0/D = 5$ $\lambda_c \simeq 0.7 - 1$ according to the short ($\gamma = 0$ in the figure) or long-range ($\gamma = 1$) nature of the phonon interaction, respectively; in this case g^2 must be at least 3.5 and 5, respectively.

The condition on small polaron formation translates into a limit on the typical size R_p of the polaron with respect to the lattice constant a. The band mass can be written in terms of the previous quantities as $m = \hbar^2/2Ja^2$ and the polaron radius is given by $R_p = \sqrt{\hbar/2m\omega_0}$; therefore it turns out that $(R_p/a) < (g/\sqrt{z\lambda_c})$, typically $R_p/a < 1$. Another feature regards the band narrowing due to lattice effects; the renormalized bandwidth W is given by $W = D exp(-g^2)$. Finally as far as the small polaron effective mass is concerned, at the formation point this is given by $m^*/m = exp(g_c^2)$ and therefore it can be very large depending on g_c^2 and consequently on the adiabatic ratio; the bandwidth in turn must allow the formation of lattice polaron bands [AlexMott94b].

2.6.2 Large (bi-)polarons

The large polaron concepts was first introduced by Fröhlich [Froh50], Landau and Pekar [LandauPekar46] and then Feynman and coworkers [Feynman62]. Large polarons in high-temperature superconductivity have been considered by Emin [Emin89], Verbist and Devreese [Devreese91], Adamoski [Adamowski89],

Bassani and coworkers [Bassani91] and Iadonisi and coworkers [Cataud91,Cata-ud92]. Large polarons are electrons which interact with the lattice over many lattice sites allowing much smaller effective masses. In the usual treatment, the formation of a large polaron is characterized by a rigid band shift due to the self-energy correction and by a correspondent increase in the effective car-rier mass. It is usual to introduce the adimensional Fröhlich electron-lattice coupling constant

$$\alpha = \frac{e^2}{\hbar\epsilon_\infty}(1-\eta)\sqrt{\frac{m}{2\hbar\omega_l}} \, ,$$

where e and m are the electron charge and band mass respectively and ω_l is the frequency of the longitudinal optical phonon involved in the lattice po-larization. Note that with respect to the coupling g^2 introduced in the small polaron case, one has $\alpha = 2g^2$. Depending on the α value electrons can be delocalized in an intermediate coupling regime ($\alpha < 10$ [Kuper63,Iadonisi84]), the motion being always coherent. At variance with the small polaron case, the polaron level shift $E_p = \alpha\hbar\omega_l$ is smaller than the half-bandwidth; this is connected to the fact that the large polaron radius is bounded from below, because it must be larger than the lattice constant.

Large bipolarons form when the coupling constant α is larger than about 4 in two dimensions and about 6 in three, as has been calculated independently and with different techniques by Bassani and coworkers [Bassani91] and De-vreese and coworkers [Devreese91]. In correspondence with and depending on these critical α values, the ionicity parameter $\eta = \epsilon_\infty/\epsilon_0$ is bounded from above.

The physics is a bit different when screening effects induced by doping are included. As far as bipolaron formation is concerned, the arising cooperation among long range excitations (phonons and plasmons) at first enhances and finally screens the bipolaron binding energy, making the ionicity parameter less relevant. Concerning the more basic polaron problem, striking conse-quences arise in relation to the self-trapping. In fact using a path integral technique it has been shown [Iadonisi96a] that for any value of α a critical density n_c exists, such that whenever $n > n_c$ the self-energy in the interme-diate coupling regime is lower than that in the strong coupling Landau and Pekar [LandauPekar46] case. This means that dynamical screening effects weaken the effective electron-phonon coupling therefore preventing polaron lo-calization. Furthermore, the interplay between α and the plasmon-phonon coupling strength $\lambda_{ph-pl} = \omega_p/\sqrt{\epsilon_\infty}\omega_l$ (see chapter 1) drives the adiabatic-ity parameter ω_0/E_F. The treatment of phonons and plasmon on the same footing brings about a reduction of the effective electron-phonon coupling α upon doping and therefore reduces the adiabaticity parameter as far as the density is increased. The previous comment will be considered in more detail in Sec. 5.3. We may then comment that small polaron theory is more accurate whenever the polaron size is comparable to the lattice site, whereas large plasma–polaron theory more correctly takes into account screening effects due to electron doping. Furthermore, in view of their relatively small mass values and intrinsic delocalization properties large polarons and bipolarons appear to be more appealing in connection with high-temperature superconductivity.

A final remark should be made. Polaron radii in high-T_c cuprates are of the order of one or two lattice constants and therefore the large polaron approach is on the limit of applicability. In particular the details of the band structure would be relevant. Concerning the latter point, it is worthwhile to note that the accuracy of the large polaron approach when the polaron extends over a lattice site can be in principle improved lifting the effective mass approximation and including the band structure profile in the kinetic energy term.

In the next chapters the physics of the plasmon–bipolaron model is developed; in particular the ground state as well as the thermodynamic properties of a system of electrons coupled to optical phonons are calculated.

Chapter 3

The plasmon–bipolaron model

Consider a cubic polar material in which the electron-optical phonon interaction is relevant and additional carriers can be introduced in the compound as a consequence of a given doping mechanism. The material we have in mind has a small r_s value because it is very ionic and therefore we may describe the correlation effects within the Random Phase Approximation [Pines63]. We wish to study the physics of a test charge added to the system described above and then consider the case of *two* additional charges of the same sign.

3.1 Equations of motion

The approach is based on a set of equations of motion which describe the dynamics of the optical phonons in the Fröhlich [Froh50] scheme and that of the electrons in the Single-Plasmon-Pole-Approximation (SPPA) [Barton71]. As far as the phonons are concerned we introduce the classical quantity $\vec{W}(\vec{r}, t)$ which represents the relative ionic displacement in the elementary lattice cell. The classical equation of motion for the spatial Fourier transform $\vec{W}_{\vec{k}}(t)$ is given by

$$\frac{d^2}{dt^2}\vec{W}_{\vec{k}}(t) = b_{11}\vec{W}_{\vec{k}}(t) + b_{12}\vec{E}_{\vec{k}} , \qquad (3.1)$$

where the constants b_{11} and b_{12} are to be determined and $\vec{E}_{\vec{k}}$ is the spatial Fourier transform of the total electric field. We now introduce the ionic polarization $\vec{P}(\vec{r}, t)$; its spatial Fourier transform $\vec{P}_{\vec{k}}(t)$ is related to the ionic displacement and to the total electric field by the constitutive relation

$$\vec{P}_{\vec{k}}(t) = b_{21}\vec{W}_{\vec{k}}(t) + b_{22}\vec{E}_{\vec{k}} , \qquad (3.2)$$

where the constants b_{21} and b_{22} are to be determined. It can be shown that $b_{12} = b_{21}$ [Evrard72].

As far as the plasmons are concerned, the effect on the dynamics of the external electron is taken into account through the electron density fluctuation

$\rho(\vec{r}, t)$ which characterizes the collective excitations of the electrical charges. In the SPPA the spatial Fourier transform of the density fluctuation satisfies the equation

$$\frac{d^2}{dt^2}\rho_{\vec{k}} = -\frac{\omega_p^2}{4\pi}i\vec{k}\cdot\vec{E}_{\vec{k}} - (\omega_k^2 - \omega_p^2)\rho_{\vec{k}} , \tag{3.3}$$

where ω_k is the k dependent plasma frequency ($\omega_p = \omega_{k=0}$), n is the average electron density and m and e are the electron band mass and charge, respectively. The coupling between the ionic motion and the electron density fluctuations is given by the total electric field $\vec{E}_{\vec{k}}$. This field must satisfy the Maxwell equations

$$i\vec{k}\cdot\vec{E}_{\vec{k}} = 4\pi\left(\rho_{\vec{k}} - i\vec{k}\cdot\vec{P}_{\vec{k}} - e\exp(-i\vec{k}\cdot\vec{r})\right) . \tag{3.4}$$

$$\vec{k}\times\vec{E}_{\vec{k}} = 0 \tag{3.5}$$

Eq.(3.4) is the Gauss law, in which the total electron density is given by the sum of the electron density fluctuation, the charge density arising from the ionic polarization and from the external electron in position $\vec{r}$. Eq.(3.5) assumes that the components of the total electric field are only along the wavevector direction. This assumption fixes $b_{11} = -\omega_t^2$, where ω_t is the frequency of the transverse optical phonon. Moreover, ions cannot follow the driving field when the latter has a frequency larger than the phonon frequency; therefore $\vec{w}_{\vec{k}} = 0$ and consequently $b_{22} = (\epsilon_\infty - 1)/(4\pi)$ where ϵ_∞ is the background high frequency dielectric constant.

From equations (3.2)– (3.5), it is possible to calculate $\vec{k}\cdot\vec{P}_{\vec{k}}$ and $\vec{k}\cdot\vec{E}_{\vec{k}}$ as a function of $\vec{k}\cdot\vec{w}_{\vec{k}}$ and $\rho_{\vec{k}}$. The direct substitution in equations (3.1) and (3.3) yields the equations of motion for the quantities $u_{\vec{k}} = \vec{k}\cdot\vec{w}_{\vec{k}}/k$ and $z_{\vec{k}} = \rho_{\vec{k}}/(ik)$:

$$\frac{d^2 u_{\vec{k}}}{dt^2} = -\omega_l^2 u_{\vec{k}} + \frac{4\pi b_{12}}{\epsilon_\infty}z_{\vec{k}} - \frac{4\pi b_{12}e}{ik\epsilon_\infty}\exp(-i\vec{k}\cdot\vec{r}) \tag{3.6}$$

$$\frac{d^2 z_{\vec{k}}}{dt^2} = -\frac{\omega_k^2}{\epsilon_\infty}z_{\vec{k}} + \frac{\omega_p^2 b_{12}}{\epsilon_\infty}u_{\vec{k}} - \frac{\omega_p^2 e}{ik\epsilon_\infty}\exp(-i\vec{k}\cdot\vec{r}) , \tag{3.7}$$

where ω_l is the longitudinal optical frequency and $b_{12} = [\epsilon_\infty(\omega_l^2 - \omega_t^2)/4\pi]^{1/2}$. The hamiltonian which gives the equations of motion (3.6) and (3.7) and the equation of motion for the electron can be quantized. The result is

$$H_{1e} = \sum_{\mathbf{k}}\left[\hbar\omega_l(a_{\mathbf{k}}^+ a_{\mathbf{k}} + \frac{1}{2}) + \frac{\hbar\omega_k}{\sqrt{\epsilon_\infty}}(b_{\mathbf{k}}^+ b_{\mathbf{k}} + \frac{1}{2}) + Z_k(a_{\mathbf{k}}^+ + a_{-\vec{k}})(b_{-\vec{k}}^\dagger - b_{\mathbf{k}})\right.$$
$$\left. + \left(V_{\mathbf{k}}e^{i\vec{k}\cdot\vec{r}}a_{\mathbf{k}} + h.c.\right) + \left(U_{\mathbf{k}}e^{i\vec{k}\cdot\vec{r}}b_{\mathbf{k}} + h.c.\right)\right] + \frac{p^2}{2m} , \tag{3.8}$$

where

$$Z_k = -\frac{i\hbar}{2}\left[\frac{\omega_p^2\omega_l}{\omega_k\sqrt{\epsilon_\infty}}(1-\eta)\right]^{1/2}$$

$$V_k = -\frac{1}{k}\left[\frac{2\pi e^2\omega_l\hbar}{V\epsilon_\infty}(1-\eta)\right]^{1/2}$$

$$U_k = -\frac{i}{k}\left[\frac{2\pi e^2\omega_p^2\hbar}{V\omega_k\epsilon_\infty^{3/2}}\right]^{1/2}$$

$$\eta = \frac{\epsilon_\infty}{\epsilon_0},$$

V being the volume. The operators a_k (a_k^+) and b_k (b_k^+) are the annihilation (creation) boson operators for phonons and plasmons, respectively.

The term containing Z_k gives the interaction between phonons and plasmons; the term driven by the coefficient V_k is the Fröhlich [Froh50] electron-phonon interaction and the term in U_k is the electron-plasmon interaction [Overhauser71]. Summarizing, equation (3.8) gives the hamiltonian of one electron, a phonon and a plasmon fields in interaction with each other.

3.2   The Hamiltonian

The hamiltonian (3.8) can be easily generalized to the case of two electrons, taking care of adding the direct Coulomb repulsion between them. The hamiltonian (3.8) then becomes:

$$
\begin{aligned}
H = &\sum_k\left[\hbar\omega_l(a_k^+a_k + \frac{1}{2}) + \hbar\frac{\omega_k}{\sqrt{\epsilon_\infty}}(b_k^+b_k + \frac{1}{2}) + Z_k(a_k^+ + a_{-k})(b_{-k}^\dagger - b_k)\right. \\
&\left. + \left(V_k\left(e^{i\vec{k}\cdot\vec{r}_1} + e^{i\vec{k}\cdot\vec{r}_2}\right)a_k + h.c.\right) + \left(U_k\left(e^{i\vec{k}\cdot\vec{r}_1} + e^{i\vec{k}\cdot\vec{r}_2}\right)b_k + h.c.\right)\right] \\
&+ \frac{p_1^2}{2m} + \frac{p_2^2}{2m} + \frac{e^2}{\epsilon_\infty|\vec{r}_1 - \vec{r}_2|}.
\end{aligned}
\tag{3.9}
$$

The Coulomb repulsion between the two external electrons is statically screened by the background high frequency dielectric constant ϵ_∞. Introducing the center of mass coordinate $\vec{R}$, the relative position $\vec{r} = \vec{r}_1 - \vec{r}_2$ and the conjugate variables $\vec{P}$ and $\vec{p}$, the hamiltonian becomes:

$$
\begin{aligned}
H = &\sum_k\left[\hbar\omega_l(a_k^+a_k + \frac{1}{2}) + \hbar\frac{\omega_k}{\sqrt{\epsilon_\infty}}(b_k^+b_k + \frac{1}{2}) + Z_k(a_k^+ + a_{-k})(b_{-k}^\dagger - b_k)\right. \\
&\left. + \left(\rho_{\vec{k}}(\vec{r})e^{i\vec{k}\cdot\vec{R}}(V_k a_k + U_k b_k) + h.c.\right)\right] + \frac{P^2}{2M} + \frac{p^2}{2\mu} + \frac{e^2}{\epsilon_\infty r},
\end{aligned}
\tag{3.10}
$$

with $M = 2m$, $\mu = m/2$ and $\rho_{\vec{k}}(\vec{r}) = 2\cos(\frac{\vec{k}\cdot\vec{r}}{2})$.

Hamiltonian (3.10) commutes with the total momentum of the system,

$$\vec{\mathcal{P}} = \vec{P} + \hbar \sum_{\vec{k}} \vec{k} \left(a_{\mathbf{k}}^+ a_{\mathbf{k}} + b_{\mathbf{k}}^+ b_{\mathbf{k}} \right) , \tag{3.11}$$

The goal is to construct the ground state of equation (3.10), taking into account the conservation law (3.11).

3.3   The Hopfield transformation

We first consider the hamiltonian

$$H_0 = \sum_{\mathbf{k}} \left[\hbar\omega_l (a_{\mathbf{k}}^+ a_{\mathbf{k}} + \frac{1}{2}) + \frac{\hbar\omega_k}{\sqrt{\epsilon_\infty}} (b_{\mathbf{k}}^+ b_{\mathbf{k}} + \frac{1}{2}) + Z_k (a_{\mathbf{k}}^+ + a_{-\vec{k}})(b_{-\vec{k}}^\dagger - b_{\mathbf{k}}) \right] . \tag{3.12}$$

The structure of Eq.(3.12) is similar to the one considered by Hopfield [Hopf58] for the polariton problem. A canonical transformation has to be found to diagonalize the hamiltonian (3.12). Given the coefficients w_{1k}, x_{1k}, y_{1k}, z_{1k}, w_{2k}, x_{2k}, y_{2k}, z_{2k}, the transformation is

$$\alpha_{\vec{k}_1} = w_{1k}a_{\mathbf{k}} + x_{1k}b_{\mathbf{k}} + y_{1k}a_{-\vec{k}}^\dagger + z_{1k}b_{-\vec{k}}^\dagger \tag{3.13}$$

$$\alpha_{\vec{k}_2} = w_{2k}a_{\mathbf{k}} + x_{2k}b_{\mathbf{k}} + y_{2k}a_{-\vec{k}}^\dagger + z_{2k}b_{-\vec{k}}^\dagger$$

$$\alpha_{-\vec{k}_1}^\dagger = y_{1k}^* a_{\mathbf{k}} + z_{1k}^* b_{\mathbf{k}} + w_{1k}^* a_{-\vec{k}}^\dagger + x_{1k}^* b_{-\vec{k}}^\dagger$$

$$\alpha_{-\vec{k}_2}^\dagger = y_{2k}^* a_{\mathbf{k}} + z_{2k}^* b_{\mathbf{k}} + w_{2k}^* a_{-\vec{k}}^\dagger + x_{2k}^* b_{-\vec{k}}^\dagger .$$

The coefficients are to be fixed in such a way to satisfy the commutation relations

$$[\alpha_{\vec{k}i}, \alpha_{\vec{q}j}^\dagger] = \delta_{ij}\delta_{\vec{k}\vec{q}} \quad , \quad [\alpha_{\vec{k}i}, \alpha_{\vec{q}j}] = 0 \tag{3.14}$$

$$\left[\alpha_{\vec{k}i}, H_0\right] = \hbar\Omega_i(\vec{k})\alpha_{\vec{k}i} \quad , \quad [\alpha_{\vec{k}i}^\dagger, H_0] = -\hbar\Omega_i(\vec{k})\alpha_{\vec{k}i}^\dagger . \tag{3.15}$$

The Ω_i in the previous formula are the new frequencies. The above problem has been solved [Mahan]; defining $\tilde{\omega}_k = \omega_k/\sqrt{\epsilon_\infty}$ one obtains, being $i = 1,2$

$$2\Omega_i^2(k) = \tilde{\omega}_k^2 + \omega_l^2 + (-1)^{i+1}\sqrt{(\tilde{\omega}_k^2 - \omega_l^2)^2 + 16\tilde{\omega}_k\omega_l|Z_k/\hbar|^2} \tag{3.16}$$

$$x_{ik} = \frac{(\tilde{\omega}_k + \Omega_i)(\omega_l^2 - \Omega_i^2)}{2T_{ki}\sqrt{\tilde{\omega}_k\Omega_i}} \tag{3.17}$$

$$y_{ik} = \frac{Z_k\tilde{\omega}_k(\omega_l - \Omega_i)}{\hbar T_{ki}\sqrt{\tilde{\omega}_k\Omega_i}} \tag{3.18}$$

$$w_{ik} = -\frac{Z_k \tilde{\omega}_k (\omega_l + \Omega_i)}{\hbar T_{ki} \sqrt{\tilde{\omega}_k \Omega_i}} \tag{3.19}$$

$$z_{ik} = \frac{(\tilde{\omega}_k - \Omega_i)(\omega_l^2 - \Omega_i^2)}{2 T_{ki} \sqrt{\tilde{\omega}_k \Omega_i}} \tag{3.20}$$

$$T_{ki} = \left[4|Z_k/\hbar|^{1/2} \tilde{\omega}_k \omega_l + (\omega_l^2 - \Omega_i^2)^2\right]^{1/2} . \tag{3.21}$$

The Hopfield canonical transformation applied to the hamiltonian (3.10) yields

$$
\begin{aligned}
H = & \sum_{\mathbf{k}} \left[\hbar\Omega_1(\alpha_{\vec{k}_1}^\dagger \alpha_{\vec{k}_1} + \frac{1}{2}) + \hbar\Omega_2(\alpha_{\vec{k}_2}^\dagger \alpha_{\vec{k}_2} + \frac{1}{2}) + \right. \\
& + \left. \left(\rho_{\vec{k}}(\vec{r}) e^{i\vec{k}\cdot\vec{R}}(\tilde{V}_k \alpha_{\vec{k}_1} + \tilde{U}_k \alpha_{\vec{k}_2}) + h.c.\right)\right] \\
& + \frac{P^2}{2M} + \frac{p^2}{2\mu} + \frac{e^2}{\epsilon_\infty r} ,
\end{aligned}
\tag{3.22}
$$

where

$$\tilde{V}_k = V_k(w_{1k}^* - y_{1k}^*) + U_k(x_{1k}^* + z_{1k}^*) \tag{3.23}$$

and

$$\tilde{U}_k = V_k(w_{2k}^* - y_{2k}^*) + U_k(x_{2k}^* + z_{2k}^*). \tag{3.24}$$

The transformed hamiltonian (3.22) refers to two charged particles interacting with two independent renormalized boson fields. The coupling of the electrons with the new fields is through $\tilde{V}_k$ and $\tilde{U}_k$, which are connected to the old coupling constants V_k and U_k.

3.4   Polaronic units

Polaronic units are $\hbar\omega_l$ for energy, the band mass m_b for the mass and the polaron radius for length, $R_p = \sqrt{\hbar/(2m_b\omega_l)}$. In these units the Frölich electron-phonon coupling constant comes out to be (remember $\eta = \epsilon_\infty/\epsilon_0$):

$$\alpha = \frac{1}{\epsilon_\infty}(1 - \eta)\frac{e^2}{2R_p}\frac{1}{\hbar\omega_l}.$$

It is useful to introduce the dimensionless parameter $\lambda = \omega_p/(\sqrt{\epsilon_\infty}\omega_l)$, The parameter λ represents the importance of the electronic effects with respect to the phonon ones and therefore drives the strength of the plasmon-phonon coupling.

3.5   Renormalized frequencies and coupling constants

The features of the new frequencies $\Omega_i(k)$ and new coupling constants $\tilde{V}_k$ and $\tilde{U}_k$ can be studied as a function of λ and introducing the dimensionless parameter $s_k = \omega_k/\omega_p$ and $\eta = \epsilon_\infty/\epsilon_0$. The new frequencies R_1 and R_2 and coupling constants F_1 and F_2 in polaronic units are given by

$$R_1^2 = \frac{1}{2}\left[\lambda^2 s^2(k) + 1 + \sqrt{(\lambda^2 s^2(k) - 1)^2 + 4\lambda^2(1-\eta)}\right] \tag{3.25}$$

$$R_2^2 = \frac{1}{2}\left[\lambda^2 s^2(k) + 1 - \sqrt{(\lambda^2 s^2(k) - 1)^2 + 4\lambda^2(1-\eta)}\right] \tag{3.26}$$

$$F_1 = -\lambda\frac{\sqrt{(1-\eta)/R_1} + (R_1^2 - 1)/\sqrt{(1-\eta)R_1}}{\sqrt{(1-R_1^2)^2 + \lambda^2(1-\eta)}} \tag{3.27}$$

$$F_2 = \lambda\frac{\sqrt{(1-\eta)/R_2} + (R_2^2 - 1)/\sqrt{(1-\eta)R_2}}{\sqrt{(1-R_2^2)^2 + \lambda^2(1-\eta)}}. \tag{3.28}$$

The relationship between R_i and Ω_i, F_i and $\tilde{V}_k$, $\tilde{U}_k$ are

$$\Omega_i(k) = R_i(k)\omega_l$$

$$\tilde{V}_k = \sqrt{\frac{4\pi\alpha}{V}}\frac{1}{k\sqrt{R_p}}F_1\hbar\omega_l$$

$$\tilde{U}_k = \sqrt{\frac{4\pi\alpha}{V}}\frac{1}{k\sqrt{R_p}}F_2\hbar\omega_l.$$

The renormalized frequencies and coupling constants are connected among themselves by a number of relations, which are reported in Appendix A.

The renormalized frequencies Ω_i in Fig. 3.1 in polaronic units as a function of λ. Let us now see the limiting behaviours of the new frequencies and coupling constants. For $\lambda \to 0$, namely small electronic density,

$$\Omega_1(k) \to \omega_l(1 + \lambda^2(1-\eta))^{1/2} \tag{3.29}$$

$$\Omega_2(k) \to \omega_l\lambda(s_k^2 - 1 + \eta)^{1/2} \tag{3.30}$$

$$\tilde{V}_k \to iV_k \tag{3.31}$$

$$\tilde{U}_k \to \eta U_k\frac{s_k^{1/2}}{(s_k^2 - 1 + \eta)^{1/4}}. \tag{3.32}$$

From the previous equations it can be seen that Ω_1 tends to the phonon longitudinal frequency and the corresponding coupling reduces to the Fröhlich term. As far as Ω_2 and $\tilde{U}_k$ are concerned, the meaning of the limits is more

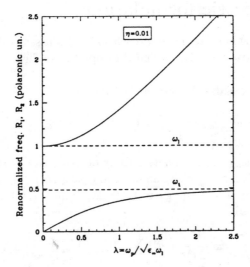

Figure 3.1: Renormalized phonon-plasmon frequencies in units of $\hbar\omega_l$ as a function of λ and for $\eta = 0.01$ (see text). The plasma frequency is supposed to be not dispersive. Note the anticrossing region ($\lambda \simeq 1$), where the two field are mostly coupled. Note that the lower branch is amplified by a factor of 5.

clear when the plasma frequency is not dispersive; in this case $\Omega_2 \to \omega_p/\sqrt{\epsilon_0}$ and $\tilde{U}_k \to U_k\eta^{3/4}$. Since the phonon frequency is larger than ω_p, it follows that even the phonons participate to the screening of the electron interactions; therefore the plasma frequency is screened by ϵ_0.

In the opposite limit $\lambda \gg 1$ we have

$$\Omega_1 \;\to\; \omega_l\left(\lambda^2 s_k^2 + \frac{1-\eta}{s_k^2}\right)^{1/2} \tag{3.33}$$

$$\Omega_2 \;\to\; \omega_l\left(1 - \frac{1-\eta}{s_k^2}\right)^{1/2} \tag{3.34}$$

$$\tilde{V}_k \;\to\; U_k \tag{3.35}$$

$$\tilde{U}_k \;\to\; -i\frac{V_k}{[1-(1-\eta)/s_k^2]^{1/4}}\left(1 - \frac{1}{s_k^2}\right) . \tag{3.36}$$

Since the phonons cannot follow any longer the electrons, Ω_1 approaches the plasma frequency ω_k (screened by ϵ_∞) and $\tilde{V}_k$ the bare electron-plasmon interaction; the limits of Ω_2 and $\tilde{U}_k$ in the case $s_k = 1$ (no dispersion) are $\Omega_2 \to \omega_t$ and $\tilde{U}_k \to 0$ and indicate a complete screening. In fact no long ranged excitations survive and therefore the longitudinal optical phonons has the frequency of the transverse phonon.

3.6   Dielectric formulation

In this section the hamiltonian approach so far discussed is shown to be equivalent to a dielectric formulation and the appropriate dielectric function of the system is given.

First, let us consider the model Hamiltonian

$$H = \sum_{\vec{k},n} \omega_{n,k} b^\dagger_{\vec{k},n} b_{\vec{k},n} + \sum_{\vec{k},n} C_{n,k} \left(b_{\vec{k},n} e^{i\vec{k}\cdot\mathbf{r}(t)} + h.c. \right) , \tag{3.37}$$

where $C_{n,k}$ and $\omega_{n,k}$ are parameters to be determined. Assume now that the external charge is moving along an arbitrary path $\mathbf{r}(t)$. The Fermi golden rule leads immediately to the energy transfer ΔE from the external charge to the boson field [Pines63].

$$\Delta E = \frac{2\pi}{\hbar} \int_0^\infty d\omega \frac{V}{(2\pi)^3} \int d^3k \sum_n \omega |C_{n,k}|^2 |\rho_{ex}(\vec{k},\omega)|^2 \delta(\omega_{n,k} - \omega) , \tag{3.38}$$

where

$$\rho_{ex}(\vec{k},\omega) = \int_{-\infty}^\infty dt e^{i(\vec{k}\cdot\mathbf{r}(t) - \omega t)} .$$

Let us now calculate the same quantity ΔE starting from the Maxwell equations and considering a medium characterized by a dielectric function $\epsilon(k,\omega)$. We obtain

$$\Delta E = 8\pi e^2 \int \frac{d^3k}{(2\pi)^3} \int_0^\infty \frac{d\omega}{2\pi} \frac{\omega}{k^2} |\rho_{ex}(\vec{k},\omega)|^2 Im \left[-\frac{1}{\epsilon(k,\omega)} \right] . \tag{3.39}$$

Comparing equations (3.38) and (3.39) one obtains

$$\sum_n |C_{n,k}|^2 \delta(\omega_{n,k} - \omega) = \frac{4e^2\hbar}{Vk^2} Im \left[-\frac{1}{\epsilon(k,\omega)} \right] . \tag{3.40}$$

It must be noted that $\epsilon(k,\omega)$ does not include the effects of the external particle.

As an example of the usefulness of Eq.(3.40) let us consider the case of an electron in a polar material. If the material is described by the approximate dielectric function [Mahan]

$$\epsilon(k,\omega) = \epsilon_\infty \left[1 - \frac{\omega_l^2 - \omega_t^2}{\omega^2 - \omega_t^2} \right] , \tag{3.41}$$

it is easy to show that

$$Im \left[-\frac{1}{\epsilon(k,\omega)} \right] = \frac{\pi}{2} \frac{\omega_l^2 - \omega_t^2}{\epsilon_\infty \omega} \delta(\omega - \omega_l) .$$

From Eq.(3.40) the usual electron-phonon Fröhlich coupling constant can be recovered. The same scheme can be used to derive the electron-plasmon coupling constant within the plasmon-pole approximation for the electron gas dielectric function.

The case we are interested in is the interaction of an external particle with a bath of phonons and plasmons. A widely used approximation for the dielectric function [Mahan] is

$$\epsilon(k, \omega) = \epsilon_\infty \left[1 - \frac{\omega_l^2 - \omega_t^2}{\omega^2 - \omega_t^2} - \frac{\tilde{\omega}_p^{\,2}}{\omega^2 + \tilde{\omega}_p^{\,2} - \omega^2(k)} \right], \tag{3.42}$$

where $\omega(k)$ is the plasmon dispersion and $\tilde{\omega}_p = \omega_p/\sqrt{\epsilon_\infty}$. The dielectric function (3.42) is the sum of three independent contributions to the screening: the background, the electrons and the phonons, electrons and phonons being treated on the same footing. Although the sum of the three independent terms would appear too rough, however it can be shown that this does provide a good description of the interaction among the different modes as long as they have *longitudinal* electric fields.

The zeros of $\epsilon(k, \omega)$ give the eigen-frequencies Ω_i of the coupled phonon-plasmon system; they are the same which have been found in Eq. (3.16). It is convenient to rewrite Eq.(3.42) in the form

$$\frac{1}{\epsilon(k, \omega)} = \frac{1}{\epsilon_\infty} \left[1 + \frac{R_1(k)}{\omega^2 - \Omega_1^2(k)} + \frac{R_2(k)}{\omega^2 - \Omega_2^2(k)} \right], \tag{3.43}$$

where

$$R_i(k) = (-1)^i \frac{\Omega_i^2(\tilde{\omega}_p^{\,2} + \omega_l^2 - \omega_t^2) - \Omega_1^2\Omega_2^2 - \omega_t^2(\tilde{\omega}_p^{\,2} - \omega_k^2)}{\Omega_2^2 - \Omega_1^2} \quad i = 1, 2. \tag{3.44}$$

Thus one easily obtains

$$Im\left[-\frac{1}{\epsilon(k, \omega)} \right] = \frac{\pi}{2\omega\epsilon_\infty} [R_1(k)\delta(\omega - \Omega_1(k)) + R_2(k)\delta(\omega - \Omega_2(k))]. \tag{3.45}$$

From Eq.(3.40) the coupling coefficients are

$$|C_{i,k}|^2 = \frac{2\pi e^2 \hbar}{V k^2 \epsilon_\infty} \frac{R_i(k)}{\Omega_i(k)}. \tag{3.46}$$

It is only a matter of lengthy algebra to show that these coupling coefficients are identical to the ones derived in the previous section. The reason of this identity lies on the internal consistency between the model dielectric function of Eq.(3.42) and the equations of motion (3.1) and (3.3).

The equivalence just shown could allow the derivation of more realistic coupling coefficients between the external charge and the renormalized fields, provided that improved total dielectric functions are used. For instance the

Singwi-Tosi-Land-Sjölander [STLS68] model dielectric function could be used for the electronic part; in this case short range correlation effects would be carefully taken into account through the static local field factor.

A further interesting point is that the knowledge of the dielectric function for the free electrons $\epsilon_{el}(k,\omega)$ allows the calculation of the plasmon dispersion relation. In fact, neglecting the ionic contribution in Eq.(3.42), one obtains

$$\omega^2(k) = \tilde{\omega}_p{}^2 \frac{\epsilon_{el}(k,0)}{\epsilon_{el}(k,0) - \epsilon_\infty} . \tag{3.47}$$

It is worth to point out that the two equivalent formulation of the problem have different advantages. On the one hand the dielectric formulation summarized in Eq.(3.42) provides more reliable renormalized frequencies Ω_i and coupling coefficients whenever a better model dielectric function is available; on the other hand the hamiltonian formulation allows to construct reliable PP and BPP states and effective electron-electron potentials.

As a final comment, we wish to point out that the dielectric function (3.42) has been necessary in order to fit the reflectivity data by Gajic et al. [Gajic92], therefore suggesting the excistence of plasmon phonon coupling at least in $YBa_2Cu_3O_{7-\delta}$. It would be interesting to fit such experimental data (reflectivity as well as Raman measurements) on a same sample at different doping levels; this would give the renormalized frequencies displayed in Fig. 3.1. Such analysis has been beautifully performed on heavily doped GaAs in the early Sixthies upon the theoretical prediction by Varga [Varga65] and the pioneering experimental and theoretical work by Mooradian and Wright [Mooradian66,Mooradian72].

3.7 A note on the application of the model to other systems

In this section we wish to shortly comment on the applicability of the scheme described so far to other systems than high-temperature superconductors. First, it is evident that the present model applies in general to heavily doped polar semiconductors. It gives indeed the possibility to study a (relatively) low density electron gas at weak coupling. As already remarked in the introductory chapter, the coupling strength in an electron gas is usually defined introducing the adimensional parameter $r_s = r_0/a_0$, r_0 being the average interparticle distance and a_0 being the effective Bohr radius. The former is related to the particle density by $r_0 = (4\pi n/3)^{-1/3}$ whereas the latter is connected to the static dielectric constant ϵ_0 and to the effective mass m by $a_0 = \hbar^2\epsilon_0/me^2$ (e is the electron charge). In particular r_s turns out to be the ratio of the average Coulomb energy to the kinetic energy. In heavily doped polar semiconductors ϵ_0 can be very large resulting in r_s values of the order of 1.

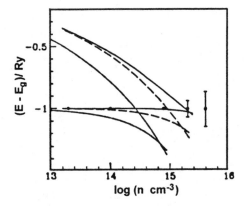

Figure 3.2: Self-energy and 1s exciton energy for GaAs as a function of the photoexcited carrier density n. Energies are in units of the effective Rydberg and E_g is the gap at $n = 0$. Dots with error bars are the experimental data after Ref. [Fehr85]. Full lines are calculated after Ref. [Ninno94]. The crossing between the two lines gives the density at which the exciton breaks up merging into the continuum (exciton bleaching) in two different approximations.

   The present picture with obvious modifications has been applied to the problem of exciton formation and exciton bleaching in strongly photoexcited semiconductors [Ninno94], [Capone94]. A good description of screening effects on the exciton binding energy has been obtained. A very striking evidence of such screening effects comes from luminescence experiments; here a pinning of the exciton line to a constant energy is observed, irrespective of the photoexcitation level. This happens because the decrease of the exciton binding energy due to the screening effects of the electron and hole plasma is compensated by the renormalization of the semiconductor energy gap, until exciton bleaching is reached. From a theoretical point of view, sophisticated many-particle techniques have been used to account for the experimental findings. The simpler one-particle picture discussed above can give a satisfactory explanation of experimental data. This has been worked out [Ninno94] and the comparison with luminescence data is displayed in Fig. 3.2. In the adopted hamiltonian description an electron and a hole interact with each other through an attractive Coulomb potential and interact with the two component plasmon field due to charge fluctuations of the other electrons and holes.

Chapter 4

The intermediate coupling regime

Let us turn back to the hamiltonian approach. A variational solution of the problem is possible within the Lee, Low and Pines (LLP) [LLP53] scheme. Such a variational procedure has been successfully applied to the large polaron problem in the intermediate coupling regime for the electron-phonon interaction, where the perturbative method is no longer suitable. LLP scheme has been shown to be valid for values of the adimensional Fröhlich coupling constant α up to 10 about [Iadonisi84]. LLP procedure uses the conservation law for the total momentum (i.e. of the two electrons plus the phonon and the plasmon fields) in order to get rid of the electron coordinates; the hamiltonian is then reduced to the hamiltonian for *two* (in our case) displaced harmonic oscillators, namely the phonon and the plasmon fields. Then a further canonical transformation is applied, which corresponds to choose the trial wave function as a product of coherent states for phonons and plasmons times an envelope function for the two electrons. This reduces the hamiltonian in a diagonal form for the plasmon and phonon operators. Finally the total energy can be calculated and minimized.

4.1   Canonical transformations

The conservation law for the total momentum (Eq.(3.11)) is used to eliminate the center of mass coordinates through the unitary transformation

$$U = \exp\left[i \left(\vec{Q} - \sum_{\mathbf{k}} \vec{k} \alpha_{\vec{k}_1}^{\dagger} \alpha_{\vec{k}_1} - \sum_{\mathbf{k}} \vec{k} \alpha_{\vec{k}_2}^{\dagger} \alpha_{\vec{k}_2} \right) \cdot \vec{R} \right] . \tag{4.1}$$

The transformed hamiltonian reads

$$\tilde{H} = U^{-1} H U = \frac{\hbar^2}{2M} \left(\vec{Q} - \sum_{\mathbf{k}} \vec{k} \alpha_{\vec{k}_1}^{\dagger} \alpha_{\vec{k}_1} - \sum_{\mathbf{k}} \vec{k} \alpha_{\vec{k}_2}^{\dagger} \alpha_{\vec{k}_2} \right)^2 + \tag{4.2}$$

$$\frac{p^2}{2\mu} + \frac{e^2}{\epsilon_\infty r} + \sum_k \left[\hbar\Omega_1(\alpha^\dagger_{\vec{k}_1}\alpha_{\vec{k}_1} + \frac{1}{2}) \right.$$

$$\left. + \hbar\Omega_2(\alpha^\dagger_{\vec{k}_2}\alpha_{\vec{k}_2} + \frac{1}{2}) + (\rho_{\vec{k}}(\vec{r})(\tilde{V}_k\alpha_{\vec{k}_1} + \tilde{U}_k\alpha_{\vec{k}_2}) + h.c.) \right] ,$$

where $\hbar\vec{Q}$ is the eigenvalue of $\vec{P}$.

The hamiltonian (4.2) is solved following a variational procedure. The trial ground state $|\psi\rangle$ is chosen to be the product of a part concerning the renormalized plasmon-phonon fields $|PlPh\rangle$ times a pair envelope function $\phi(r)$, The plasmon-phonon part is taken to be a coherent state, namely a linear coherent superposition of stationary states of the two renormalized harmonic oscillators (phonons and plasmons) [Cohen]. $|PlPh\rangle$ can be written as $|PlPh\rangle = U_1(\vec{r})U_2(\vec{r})|0>$, where $|0\rangle$ is the vacuum of the renomalized phonon-plasmon creation and destruction operators $\alpha_{\vec{k}_1}$ and $\alpha_{\vec{k}_2}$ and the operators U_1 and U_2 are given by

$$U_1(\vec{r}) = \exp\left[\sum_k \left(f_\mathbf{k}(\vec{r})\alpha_{\vec{k}_1} - f^*_\mathbf{k}(\vec{r})\alpha^\dagger_{\vec{k}_1} \right) \right] \tag{4.3}$$

$$U_2(\vec{r}) = \exp\left[\sum_k \left(g_\mathbf{k}(\vec{r})\alpha_{\vec{k}_2} - g^*_\mathbf{k}(\vec{r})\alpha^\dagger_{\vec{k}_2} \right) \right] ; \tag{4.4}$$

the functions f and g have to be determined. They are the distribution functions of the two renormalized fields. It is noteworthy to point out that they do depend on the relative coordinate r and therefore self-consistently drive the dynamics of the system.

The envelope function ϕ is chosen as a $1s$ hydrogenic-like wave function

$$\phi(r) = \frac{(2\gamma)^{\beta+3/2}}{\sqrt{\Gamma(2\beta+3)}} r^\beta e^{-\gamma r} , \tag{4.5}$$

where γ and β are variational parameters. Eq.(4.5) contains physical information on the two particle state: it must be unlikely to have the two particles very close to each other and to have a bound state when they are very far apart. Finally,

$$|\psi\rangle = U_1(\vec{r})U_2(\vec{r})|0> \phi(r) . \tag{4.6}$$

4.2   The variational method

The set of functions ϕ, $f_\mathbf{k}$ and $g_\mathbf{k}$ are determined variationally upon minimization of the total energy

$$E_T(Q) = \langle\psi|\tilde{H}|\psi\rangle$$

$$= \langle \phi | \left[\frac{\hbar^2}{2M}(\vec{Q} - \vec{K}_1 - \vec{K}_2)^2 + \frac{1}{2\mu}(\vec{j}_1 + \vec{j}_2)^2 + \right. \tag{4.7}$$

$$\frac{p^2}{2\mu} + \frac{e^2}{\epsilon_\infty r} + \frac{\hbar^2}{2\mu} \sum_{\mathbf{k}} \left(|\nabla f_{\mathbf{k}}|^2 + |\nabla g_{\mathbf{k}}|^2 \right) +$$

$$\frac{\hbar^2}{2M} \sum_{\mathbf{k}} \left(k^2 |f_{\mathbf{k}}|^2 + k^2 |g_{\mathbf{k}}|^2 \right) +$$

$$\sum_{\mathbf{k}} \left(\hbar \Omega_1 |f_{\mathbf{k}}|^2 + \hbar \Omega_2 |g_{\mathbf{k}}|^2 \right) -$$

$$\left. \sum_{\mathbf{k}} \left(\rho_k (\tilde{V}_k f_{\mathbf{k}}^* + \tilde{U}_k g_{\mathbf{k}}^* + c.c.) \right) \right] | \phi \rangle ,$$

where the following quantities have been defined:

$$\vec{K}_1 = \sum_{\mathbf{k}} \vec{k} |f_{\mathbf{k}}|^2 \tag{4.8}$$

$$\vec{K}_2 = \sum_{\mathbf{k}} \vec{k} |g_{\mathbf{k}}|^2 \tag{4.9}$$

$$\vec{j}_1 = \frac{\hbar}{2i} \sum_{\mathbf{k}} [f_{\mathbf{k}} \nabla f_{\mathbf{k}}^* - f_{\mathbf{k}}^* \nabla f_{\mathbf{k}}] \tag{4.10}$$

$$\vec{j}_2 = \frac{\hbar}{2i} \sum_{\mathbf{k}} [g_{\mathbf{k}} \nabla g_{\mathbf{k}}^* - g_{\mathbf{k}}^* \nabla g_{\mathbf{k}}] \tag{4.11}$$

The $\vec{K}_i$s have the meaning of the average momentum of the two renormalized boson fields.

The functional variation of $E_T(Q)$ with respect to $f_{\mathbf{k}}^*$ and $g_{\mathbf{k}}^*$ gives the following differential equations [Bassani91,Quattropani79]

$$- \frac{\hbar^2}{2\mu} \left(\nabla^2 f_{\mathbf{k}} + \frac{1}{\phi^2} \nabla \phi^2 \cdot \nabla f_{\mathbf{k}} \right) \tag{4.12}$$

$$+ \left(\hbar \Omega_1 + \frac{\hbar^2 k^2}{2M} - \frac{\hbar^2}{M} \vec{k} \cdot (\vec{Q} - \vec{K}_1 - \vec{K}_2) \right) f_{\mathbf{k}}$$

$$- \frac{\hbar}{2i\mu} \frac{1}{\phi^2} \left(\nabla \cdot \left((\vec{j}_1 + \vec{j}_2) \phi^2 f_{\mathbf{k}} \right) + \phi^2 (\vec{j}_1 + \vec{j}_2) \cdot \nabla f_{\mathbf{k}} \right)$$

$$= \tilde{V}_k \rho_k \tag{4.13}$$

and

$$- \frac{\hbar^2}{2\mu} \left(\nabla^2 g_{\mathbf{k}} + \frac{1}{\phi^2} \nabla \phi^2 \cdot \nabla g_{\mathbf{k}} \right) \tag{4.14}$$

$$+ \left(\hbar \Omega_2 + \frac{\hbar^2 k^2}{2M} - \frac{\hbar^2}{M} \vec{k} \cdot (\vec{Q} - \vec{K}_1 - \vec{K}_2) \right) g_{\mathbf{k}}$$

$$- \frac{\hbar}{2i\mu} \frac{1}{\phi^2} \left(\nabla \cdot \left((\vec{j}_1 + \vec{j}_2) \phi^2 g_{\mathbf{k}} \right) + \phi^2 (\vec{j}_1 + \vec{j}_2) \nabla g_{\mathbf{k}} \right)$$

$$= \tilde{U}_k \rho_k . \tag{4.15}$$

These differential equations are non linear and coupled through $\vec{j}_{1,2}$ and $\vec{K}_{1,2}$. Setting $Q = 0$ and assuming $\vec{j}_1 = \vec{j}_2 = \vec{K}_1 = \vec{K}_2 = 0$ it can be seen that the solutions f_k and g_k are consistent with this assumption. However, this is not true when $Q \neq 0$. In this case we use an approximation already tested in polaron theory in order to linearize the equations. The approximation consists in the substitution of $\vec{K}_1$ and $\vec{K}_2$ with their average values on ϕ

$$\langle \phi | \vec{K}_{1,2} | \phi \rangle = \sigma_{1,2} \vec{Q} \,. \tag{4.16}$$

The assumed proportionality to $\vec{Q}$ allows the self-consistent evaluation of the constants $\sigma_{1,2}$. The contributions from $\vec{j}_{1,2}$ are a posteriori verified to be negligible [Bassani91].

The solutions of the equations (4.12) and (4.14) can be searched in form of series of spherical harmonics [Iadonisi89,Strinati87]

$$f_k(\vec{r}) = \left(\frac{(2\pi)^3}{V} \right)^{1/2} \frac{1}{k} \sum_{l,m} \tilde{f}_{\vec{k},l}(r) Y^*_{l,m}(\omega_{\vec{k}}) Y_{l,m}(\omega_{\vec{r}}) \tag{4.17}$$

$$g_k(\vec{r}) = \left(\frac{(2\pi)^3}{V} \right)^{1/2} \frac{1}{k} \sum_{l,m} \tilde{g}_{\vec{k},l}(r) Y^*_{l,m}(\omega_{\vec{k}}) Y_{l,m}(\omega_{\vec{r}}) \,. \tag{4.18}$$

Within the discussed approximations it is possible to find an exact solution of both Eqs. (4.12) and (4.14):

$$\tilde{f}_{\vec{k},l} = \frac{2\mu}{\hbar^2} \left(\frac{V}{(2\pi)^3} \right)^{1/2} \tilde{V}_k 4\pi i^l \xi_1^{\zeta-2} r^\zeta e^{-\xi_1 r/2} \frac{\Gamma(a_1)}{\Gamma(b)} \tag{4.19}$$

$$\left[\Phi(a_1, b, \xi_1 r) \int_{\xi_1 r}^{\infty} dt\, t^{b-\zeta} e^{-t(1-\epsilon_1/2\xi_1)} j_l(kt/2\xi_1)(1 + (-1)^l) \Psi(a_1, b, \xi_1 t) \right.$$

$$\left. \Psi(a_1, b, \xi_1 r) \int_0^{\xi_1 r} dt\, t^{b-\zeta} e^{-t(1-\epsilon_1/2\xi_1)} j_l(kt/2\xi_1)(1 + (-1)^l) \Phi(a_1, b, \xi_1 t) \right] \,.$$

The various defined quantities are summurized below:

$$\zeta = \frac{[(2\beta+1)^2 + 4l(l+1)]^{1/2} - (2\beta+1)}{2} \tag{4.20}$$

$$\xi_1 = \left[4\gamma^2 + \frac{8\mu}{\hbar^2} \left(\hbar\Omega_1 + \frac{\hbar^2 k^2}{2M} - \frac{\hbar^2}{M} \vec{k} \cdot \vec{Q}(1-\sigma) \right) \right]^{1/2} \tag{4.21}$$

$$\epsilon_1 = \xi_1 - 2\gamma \tag{4.22}$$

$$b = 2(\beta + 1 + \zeta) \tag{4.23}$$

$$a_1 = \zeta + \frac{\epsilon_1}{\xi_1}(\beta + 1) \tag{4.24}$$

$$\sigma = \sigma_1 + \sigma_2 \,. \tag{4.25}$$

The solution for $g_{\vec{k},l}$ is identical in form to Eq. (4.19) and is obtained simply changing the subscript 1 with 2.

We may now solve Eqs.(4.19) and its $g_{\mathbf{k}}$ analog self-consistently together with the minimization of the total energy (4.7) with respect to the variational parameters of the envelope function. This allows a detailed study of the ground state properties of the system, which will be presented in the next two chapters.

Chapter 5

The effective potential for biplasmapolarons

The present chapter is devoted to the study of the effective electron–electron potential which can be defined within the plasmapolaron model. The properties of the self–energy are also analyzed.

5.1 Calculation of the effective potential

The minimization of equation (4.7) with respect to the state ϕ shows that ϕ is an eigenstate of a Schrödinger-like equation for a particle of mass μ with the effective electron-electron potential which depends self-consistently on the pair wave-function. After some lengthy algebra it can be shown that the effective potential V_{eff} is given by

$$
\begin{aligned}
V_{eff}(r) \;=\; & \frac{e^2}{\epsilon_\infty r} + \frac{\hbar^2}{2\mu} \sum_{\mathbf{k}} \left(|\nabla_{\vec{r}} f_{\mathbf{k}}(\vec{r})|^2 + |\nabla_{\vec{r}} g_{\mathbf{k}}(\vec{r})|^2 \right) \\
& + \frac{\hbar^2}{2M} \sum_{\mathbf{k}} \left(k^2 |f_{\mathbf{k}}(\vec{r})|^2 + k^2 |g_{\mathbf{k}}(\vec{r})|^2 \right) \\
& + \sum_{\mathbf{k}} \left(\hbar\Omega_1 |f_{\mathbf{k}}(\vec{r})|^2 + \hbar\Omega_2 |g_{\mathbf{k}}(\vec{r})|^2 \right) \\
& - \sum_{\mathbf{k}} \left(\rho_k (\tilde{V}_k f_{\mathbf{k}}(\vec{r}) + \tilde{U}_k(\vec{r}) g_{\mathbf{k}}) + c.c. \right) \; .
\end{aligned}
\tag{5.1}
$$

This quantity has very interesting properties. We consider in the following only the case $Q = 0$. As a first step we proceed in the analysis of those limiting behaviours of Eq.(5.1), which reveal to be the most interesting features of the effective potential. Some comments on the detailed structure of V_{eff} in the general case are given afterwards.

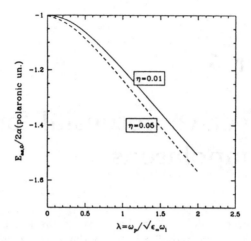

Figure 5.1: Since the self-energy of a free plasmapolaron is proportional to α, the quantity $E_{as,0}/2\alpha$ is drawn as a function of λ for $\eta = 0.01$ and 0.05. The energies are in units $\hbar\omega_l$.

5.2   The self-energy

In the limit $r \to \infty$, V_{eff} can be calculated from Eq.(5.1) inserting the corresponding asymptotic expressions for $f_{\mathbf{k}}$ and $g_{\mathbf{k}}$:

$$f_{\mathbf{k}}^{\infty}(\vec{r}) = \frac{2\mu}{\hbar^2}\tilde{V}_k \frac{e^{i\frac{\vec{k}\cdot\vec{r}}{2}} + e^{-i\frac{\vec{k}\cdot\vec{r}}{2}}}{\left[\gamma^2 + \frac{2\mu}{\hbar}\Omega_1 + \frac{1}{2}k^2\right]^{\frac{1}{2}}} \tag{5.2}$$

$$g_{\mathbf{k}}^{\infty}(\vec{r}) = \frac{2\mu}{\hbar^2}\tilde{U}_k \frac{e^{i\frac{\vec{k}\cdot\vec{r}}{2}} + e^{-i\frac{\vec{k}\cdot\vec{r}}{2}}}{\left[\gamma^2 + \frac{2\mu}{\hbar}\Omega_2 + \frac{1}{2}k^2\right]^{\frac{1}{2}}} . \tag{5.3}$$

The effective potential obtained in this limit contains a constant term, $E_{as,0}^{(2)}$, which represents the self-energy of the pair. This conclusion is similar to that obtained by other authors for the excitons [Gay72,HaugSchmitt84]. The self-energy $E_{as,0}$ is given by:

$$E_{as,0}^{(2)} = -2\alpha\hbar\omega_l \left(\frac{F_1^2}{\sqrt{R_1}} + \frac{F_2^2}{\sqrt{R_2}} \right) . \tag{5.4}$$

R_i and F_i are the renormalized frequencies and coupling constants in polaronic units, respectively (see Eqs.(3.25-3.28)). The quantity in parenthesis is displayed in Fig. 5.1.

Let us schematize the physical meaning of the self-energy in a simple band picture (see left panel in Fig. 5.2). Consider an empty conduction band (solid line in the figure). As far as more and more electrons are put in it, the bottom rigidly shifts downward (dotted curve in the figure). At the same time the bandwidth shrinks. The shrinking corresponds to an effective mass reduced by screening effects (see Sec. 6.4). If only two particles were present, then E_{as} would include only the shift due to phonon effects and in this case the band would be the flattest possible. In general, the shift and the shrinking as well take into account phonon and plasmon effects. The previous comments yield a picture in which the external charges can be described as quasi-particles including the interaction effects due to phonons and other electrons within the already discussed approximations. In the following these quasi-particles will be referred to as *plasma-polarons* (PP); they are characterized by a free particle energy spectrum with an effective carrier mass, the energy spectrum being shifted by the self-energy term.

Actually we have derived the total energy for a pair of such quasi-particles (which we analogously call biplasmapolarons or shortly BPP). The biplasmapolaron energy ($E_{T,bpp}$ in the figure) depends on the density too. The bipolaron binding energy is therefore the difference between $E_{T,bpp}$ and the self-energy of two particles, which we have indicated as E_{as}. Both $E_{T,bpp}$ and E_{as} are measured with respect to the empty reference band. Therefore it turns out that, depending on the carrier density, the BPP state may lie below or above the renormalized bottom of the conduction band with N particles (the dotted line). In the former case (it lies below) the two external electrons would form a bound state; this situation is studied in more detail in the next chapter. In the latter case, they merge in the continuum of the band and one can ask if they may constitute a resonant state (right panel of Fig. 5.2). Sec. 5.6 is devoted to some comments on this important point.

5.2.1 Asymptotic limits of the self-energy

Eq.(5.4) can be studied in three particular cases:
a) if $\lambda \to 0$, one finds $E_{as,0} \to -2\alpha\hbar\omega_l$, which is the self-energy of two independent electrons interacting only with the phonon field (two independent polarons);
b) if $\eta \to 1$, the ionic effects in the crystal are negligible; in this case

$$E_{as,0} \to E_{pl} = -\frac{e^2}{\epsilon_\infty R_{pl}},$$

R_{pl} being the "plasmaron" radius, $R_{pl} = \sqrt{\hbar\sqrt{\epsilon_\infty}/(2m\omega_p)}$. This is the self-energy of two electrons interacting with the plasmon field in a picture where the plasmon field is treated on the same footing with respect to the Fröhlich

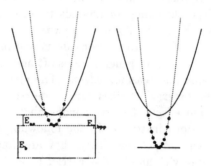

Figure 5.2: Schematic plot on the physical meaning of the self-energy. Solid lines are the reference empty band. The dotted line indicates the renormalized band. $E_{T,bpp}$ is the total BPP energy, whereas E_{as} is the self-energy of 2 PPs and E_b the BPP the binding energy.

phonon field. This value, as in the case of the polaron, is connected with the dynamics of the external electrons and plasmons;

c) if $\lambda \to \infty$, we obtain the same limit as in point b). This means that the increase in electron density screens the ionic effects; in addition the self-energy due to the plasmon field tends to $-\infty$. In Fig. 5.3 we show the quantity $\Delta = E_{as,0} - E_{pl}$ as a function of λ. Δ can be interpreted as the phonon contribution to the self-energy of the pair, including the effects due to the dynamics of the other electrons. We see that for $\lambda \leq 2$ (i.e. $n \leq 10^{20} cm^{-3}$) the dynamics of the electrons tends to partially screen the electron-phonon interaction. This fact is relevant for the biplasmapolaron formation. On the other hand, for metallic densities ($n \sim 10^{22} cm^{-3}$), the electron screening is complete, as expected.

5.3   Screening effects of electrons and phonons and the coupling regime

The introduction of screening effects in the large polaron model has striking consequences on the range of validity of the *large* polaron approach itself. The large polaron approach is indeed well defined as far as the gain in vibrational energy E_p is smaller than the electron bandwidth D. In this case, the effective

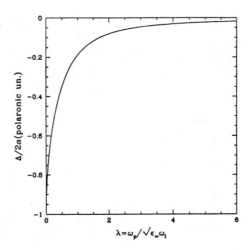

Figure 5.3: The difference of the pair self-energy in the phonon and plasmon field and that of the plasmon pair self-energy, $\Delta/2\alpha$, as a function of λ (Δ measures only the ionic contribution). We see that the phonon effects are meaningfull for electronic density less than $\lambda = 4$ (i.e. $n \sim 10^{20} cm^{-3}$).

mass approximation holds and the approach used so far is valid. The previous condition implies the existence of a critical value α_c. If however the screening effects are taken into account, the density too is important. In fact in this case the electron-phonon coupling constant decreases with increasing density. Thus let us see how the parameter E_p/D looks like. The gain in energy to be considered is the one due only to the ionic effects. This is obtained subtracting the plasmon contribution alone from half the total self-energy Eq.5.4 (we are considering only one plasmapolaron). The plasmon contribution E_{pl} is given either by the $\lambda \to \infty$ or by the $\eta \to 1$ limit of Eq.5.4, namely

$$E_{pl} = \frac{e^2}{\epsilon_\infty R_{pl}} = \frac{\sqrt{\lambda}}{1-\eta}\alpha\hbar\omega_l .$$

The difference $E_p = E_{as,0} - E_{pl}$ is the same quantity Δ which is displayed in Fig. 5.3, apart from the *alpha* scaling.

Using the same notations, the bandwidth $D \simeq E_F$ can be expressed as

$$D \simeq \left[\frac{3\pi(1-\eta)}{16}\right]^{2/3} \left(\frac{\lambda^2}{\alpha}\right)^{2/3} \hbar\omega_l .$$

The ratio E_p/D is displayed in Fig. 5.4 as a function of λ and for different values of α. It is evident that there exists a critical value λ_c such that $E_p/D < 1$

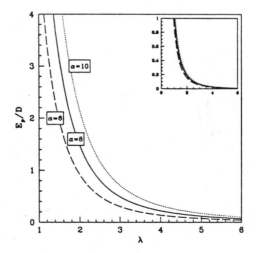

Figure 5.4: E_p/D (see text) as a function of λ for different values of α. Inset: the same quantity divided by α.

for $\lambda > \lambda_c$ and therefore the large polaron approach is meaningful. In one of the worst cases, namely $\alpha = 8$, $\lambda_c \simeq 2$ or $n \simeq 1.7 \times 10^{20}$ cm^{-3}.

The relationship between the polaron radius and the lattice parameter deserves a final comment. A direct consequence of the condition $E_p < D$ for a *traditional* large polaron is that the polaron radius turns out to be larger than the lattice constant, as expected. This can be simply derived taking into account the definition of polaron radius $R_p = \sqrt{\hbar/2m\omega_l}$ and the fact that the band mass m can be expressed in terms of the bandwidth D as

$$ m = \frac{\hbar^2 z}{Da^2} \, , $$

a being the lattice parameter and z the lattice coordination number. A further relation is needed which links E_p with some of the above quantities; such relation can be written $E_p = \alpha^* \hbar \omega_l$. In the case of the traditional large polaron $\alpha^* = \alpha$, where α has the usual definition. As far as the plasmapolaron is concerned, we can preserve the structure $\alpha^* \hbar \omega_l = e^2/2\epsilon^* R_p^*$ provided that

$$ \frac{1}{R_p^*} = \frac{f^*}{R_p} - \frac{1}{R_{pl}(1-\eta)} = \frac{1}{R_p}\left[f^* - \frac{\sqrt{\lambda}}{1-\eta}\right] \, . $$

In the previous equation

$$ f^* = \frac{F_1^2}{\sqrt{R_1}} + \frac{F_2^2}{\sqrt{R_2}} $$

and as usual

$$\frac{1}{\epsilon^*} = \frac{1}{\epsilon_\infty} - \frac{1}{\epsilon_0}.$$

The quantity f^* is once again the same as that displayed in Fig. 5.3 and precisely $f^* = \Delta/2\alpha$.

From the previous analysis we may draw two important conclusions. First, the effective plasmapolaron radius R_p^* turns out to be larger than the lattice parameter and thus the effective mass approximation is quite appropriate in a wide density range. Secondly, within the present scheme the intermediate coupling regime can be used even for large values of the electron-phonon coupling constant α, provided that the carrier density is sufficiently high. The analysis of the present section is supported by a recent study on the delocalization of self-trapped polarons due to screening effects, carried out with the Feynman path-integral technique [Iadonisi96b].

5.4   The asymptotic effective potential ($\gamma \to 0$)

The calculation of the effective potential Eq.(5.1) in the most general case is to be carried out numerically. It requires the knowledge of the parameters γ and β which minimize the total energy, all the other physical parameters (density, electron-phonon coupling constant and so forth) being fixed. As far as γ and β are different from zero, such a calculation is alternative to the procedure shown in the next chapter, where the numerical minimization of the total energy leads to the calculation of various bipolaron properties.

The situation is completely different when the minimized variational parameters vanish. In this case the bipolaron envelope function (4.6) is spatially uniform and nothing can be inferred on the bipolaron state. Indeed the bipolaron wave function factorizes in the two correspondent polaron wave functions. However we are going to show that the effective potential still retains a physically meaningful structure. When the limit $\gamma, \beta \to 0$ of the effective potential Eq.(5.1) is carried out, the effective potential takes the asymptotic form:

$$V_{as}(r) = \frac{e^2}{\epsilon_\infty r}(1-\eta)\left[\frac{F_1^2}{R_1}e^{-r\frac{\sqrt{R_1}}{R_p}} + \frac{F_2^2}{R_2}e^{-r\frac{\sqrt{R_2}}{R_p}}\right] - \qquad (5.5)$$

$$- \frac{e^2}{2\epsilon_\infty R_p}(1-\eta)\left[\frac{F_1^2}{\sqrt{R_1}}e^{-r\frac{\sqrt{R_1}}{R_p}} + \frac{F_2^2}{\sqrt{R_2}}e^{-r\frac{\sqrt{R_2}}{R_p}}\right]$$

The asymptotic effective potential as a function of the average electron-electron distance is displayed in Fig. 5.5 for $\alpha = 8$ and different densities. Fig. 5.6 shows the comparison among asymptotic effective potentials for different values of the Fröhlich coupling constant α.

It is worth to point out that even in this asymptotic limit the effective potential retains its short-range attractive character, depending on the microscopical quantities m_b, ω_l, n and on the ionicity η of the material.

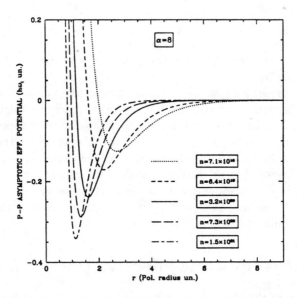

Figure 5.5: Effective asymptotic potential as function of the mean electron-electron distance r for different densities and fixed value of the Fröhlich electron-phonon coupling constant $\alpha = 8$. The effective potential is in units of $\hbar\omega_l$.

A look at Figs.5.5, 5.6 reveals that the attractive well shrinks and deepens as far as the carrier density increases and the minimum shifts towards smaller electron-electron average distances. On the other hand, the area of the attractive part gets wider with increasing the electron-phonon coupling, as is expected. The density dependence is explained, in our model, by the following picture: when the screening effects increase (for high electron densities), the induced charge and lattice polarization become fairly well localized around each electron and the attractive part of the polaron-polaron effective potential becomes narrower. Consequently the binding energy decreases (approaches zero from negative values) with increasing density.

Some meaningful limits can be recovered from Eq.(5.5), with the help of Eqs.(3.25-3.28). First, the long range tail in the $\lambda \to 0$ limit is correctly given by

$$V_{as}(\infty) \to \frac{e^2}{\epsilon_0 r}; \quad \lambda \to 0.$$

In this limit the plasmon screening effects are absent and thus the effective potential is long ranged and screened by the static dielectric constant. The latter result is well known in exciton theory, apart from an obvious minus sign in the Coulomb potential [Pollmann77,Iadonisi85]. The long range tail vanishes when $\lambda \neq 0$. The effect is of course enhanced as far as $\lambda \to \infty$ or

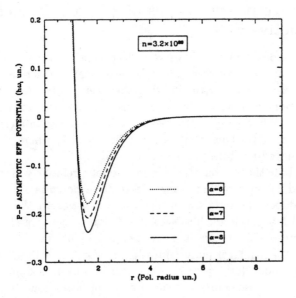

Figure 5.6: Effective asymptotic potential as function of the mean electron-electron distance r for different α values and fixed density ($n = 3.2 \times 10^{20}$ cm^{-3}). The effective potential is in units of $\hbar\omega_l$.

even $\eta \to 1$.

The remaining terms are exponentially decaying with screening constant given by

$$R_p^{-1} = \left(\frac{2m\omega_l}{\hbar}\right)^{1/2}$$

and

$$R_{pl}^{-1} = \left(\frac{2m\omega_p}{\hbar\sqrt{\epsilon_\infty}}\right)^{1/2}.$$

R_p and R_{pl} are the average polaron and "plasmaron" [Note] radius, respectively. The relative weight of these remaining terms depends on λ indeed: for $\lambda \to 0$, the screening factor is R_p^{-1} and in the opposite limit is R_{pl}^{-1}.

Along these lines, it can be shown that the whole effective potential (5.5) contains the same physics with respect to the aforementioned analysis. The differences come only from the detailed structure due to multipole terms in the phonon and plasmon distribution functions $f_{\mathbf{k}}$ and $g_{\mathbf{k}}$.

5.4.1 Relatively large values of η and existence of the attractive well

All the previous considerations have very deep consequences in the present theory. In fact, as we have seen, the existence of an attractive well in the effective potential is driven by the interplay between electronic and phononic effects. The important point to be stressed is that the attractive well exists independently of how large η is, provided that λ is larger than a critical value (which depends on α). This result is clear in Fig. 5.7, where the effective potential is displayed for different values of η and λ.

We can conclude that materials having values of η relatively large may still be characterized by such a short ranged attractive potential, provided that the material is doped. Materials with such characteristics are the alkali doped $A_x C_{60}$, where A labels the alkali species. For instance the insulating member $K_6 C_{60}$ is characterized by the ratio $\eta = \epsilon_\infty/\epsilon_0 \simeq 0.6$ [SohFink93].

Incidentally, the insulator material $K_6 C_{60}$ has very small LO-TO splittings for each phonon mode $(\omega_t^2/\omega_l^2 \simeq 1)$. The reason why η is different from 1 lies in the fact that these compounds have many optical phonons. In fact the generalized Lyddane-Sachs-Teller (LST) relation [AshMerm] tells that $\epsilon_\infty/\epsilon_0$ is given by the product of ω_t^2/ω_l^2 running over all the phonon branches. This fact can be checked just putting all the LO and TO modes of $K_6 C_{60}$ in the LST relation. This has been done using the *ab initio* calculated phonon modes by P. Giannozzi and W. Andreoni [Giannozzi96] and the agreement with the experimental value of Ref. [SohFink93] is excellent.

5.5   Polarization charges

This section is devoted to the calculation of the polarization charge distribution induced by two external charged impurities. The polarization is due to plasmon and optical phonon degrees of freedom treated on the same footing.

The density operator for the induced polarization charge can be obtained from the constitutive equations Eqs.(3.1,3.2, 3.3) together with the Maxwell Eqs.(3.4, 3.5). The second quantization in terms of the renormalized fields operator $\alpha_{\vec{k}_1}$, $\alpha_{\vec{k}_2}$ gives

$$
\rho^{tot}(k) = \sqrt{\frac{\hbar \omega_l V (1 - \eta)}{8\pi \epsilon_\infty}} ik \left[F_1(\alpha^\dagger_{-\vec{k}_1} - \alpha_{\vec{k}_1}) + F_2(\alpha^\dagger_{-\vec{k}_2} - \alpha_{\vec{k}_2}) \right] +
$$
$$
+ \; e \left(1 - \frac{1}{\epsilon_\infty} \right) \left[e^{i\vec{k}\cdot\vec{r}_0} \right] ,
$$

r_0 indicating the position of the external charge.

The limit $\gamma, \beta \to 0$ of the BPP wave function means that the envelope function $\phi(r)$ has to be put identically equal to 1; in addition $f_\mathbf{k}$ and $g_\mathbf{k}$ must

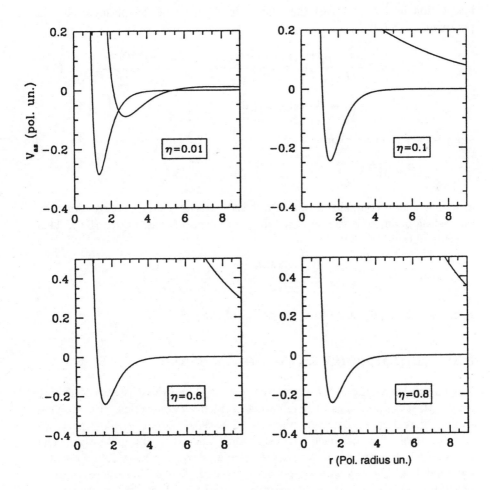

Figure 5.7: The asymptotic effective potential $V_{as}(r)$ for different values of η as indicated in figure. In each window V_{as} for two different values of λ is displayed. The upper curve corresponds to $\lambda = 0.01$ (very small density) and the lower one to $\lambda = 4$ ($n \simeq 10^{20-21}$ cm^{-3}). Note the existence of the attractive well even for large $\eta = \epsilon_\infty/\epsilon_0$ in the doped regime.

be replaced by the correspondent limits $f_{\mathbf{k}}(\gamma, \beta = 0)$ and $g_{\mathbf{k}}(\gamma, \beta = 0)$. The BPP wave function is now reduced to the wave function of one single polaron, provided that only one external charge is considered. The calculation of the expectation value of $\rho^{tot}(k)$ on the polaron wave function and then the integration in $\vec{k}$-space gives the charge distribution of the plasmon-phonon polarization cloud:

$$
\begin{aligned}
\rho(\vec{R}, \vec{r}_0) &= \frac{e(1-\eta)}{4\pi R_p^2 \epsilon_\infty} \left[F_1^2 \frac{e^{-\sqrt{R_1}|\vec{R}-\vec{r}_0|/R_p}}{|\vec{R}-\vec{r}_0|} + F_2^2 \frac{e^{-\sqrt{R_2}|\vec{R}-\vec{r}_0|/R_p}}{|\vec{R}-\vec{r}_0|} \right. \\
&+ \left. e\left(1 - \frac{1}{\epsilon_\infty}\right) \delta(\vec{R}-\vec{r}_0) \right].
\end{aligned}
\tag{5.6}
$$

It is easy to check that

$$
\int \rho(\vec{R}, \vec{r}_0) d\vec{R} = \frac{e}{\epsilon_\infty},
$$

as it must be to compensate the introduced external charge $-e/\epsilon_\infty$. Another interesting quantity is the potential due to the charge distribution (5.6). This is easily obtained to be

$$
\Phi(R) = \frac{e(1-\eta)}{\epsilon_\infty} \sum_{i=1}^{2} \frac{F_i^2}{R_i} \frac{e^{-\sqrt{R_i} R/R_p} - 1}{R} - \frac{e}{R}\left(1 - \frac{1}{\epsilon_\infty}\right),
$$

R being now the distance from the center of the cloud.

5.6   Bipolarons as resonant states

The physical meaning of the asymptotic potential and the existence of a short range attractive well poses the following problem in connection with the possibility of real space pairing. Suppose to have a system of many particles which interact via an effective short ranged and attractive potential. The question is if the system does have well-defined resonant states, namely states narrower than the resonant energy itself. This is a very difficult problem to be addressed and a large apparently not conclusive literature exists on the subject [RannRobin95,Micnas95,Randeria95].

In order to investigate the pair excitations of a Fermi gas interacting via an attractive potential $V(r)$ we study the ladder approximation to the Bethe-Salpeter integral equation for the effective two-particle interaction $\Gamma(\mathbf{p}_1, \mathbf{p}_2; -\mathbf{p}_3, \mathbf{p}_4)$ [FetterWalecka] where $\hbar\mathbf{p}_1, \hbar\mathbf{p}_2$ and $\hbar\mathbf{p}_3, \hbar\mathbf{p}_4$ are the four-momentum of the incoming and outgoing particles, respectively. This quantity is closely related to the two-particle Green's function and may be interpreted as a generalized scattering amplitude in the interacting Fermi gas. Since both total momentum $\hbar\vec{P}$ and energy, E, are conserved it is possible to introduce the center of mass momentum $\hbar\vec{P}$ and the relative momenta $\hbar\vec{p} = \frac{\hbar}{2}(\vec{p_1} - \vec{p_2})$, $\hbar\vec{p}' = \frac{\hbar}{2}(\vec{p_3} - \vec{p_4})$; then we can write

$$\Gamma(\mathbf{p_1}, \mathbf{p_2}; \mathbf{p_3}, \mathbf{p_4}) \equiv \Gamma(\vec{p}, \vec{p}'; \vec{P}, E). \tag{5.7}$$

In the following we will restrict our analysis to the effective two-particle interaction in the case of zero center of mass momentum, $\vec{P} = 0$, and on the energy shell, $|\vec{p}| = |\vec{p}'| = \sqrt{2mE}/\hbar$. Following Ref. [FetterWalecka] we introduce a pseudo-wave function $\chi(\vec{r}, \vec{p}; E)$ defined by

$$\Gamma(\vec{p}; E) = \int d\vec{r} V(r) e^{-i\vec{p}\cdot\vec{r}} \chi(\vec{r}, \vec{p}; E). \tag{5.8}$$

The Bethe-Salpeter integral equation for $\chi(\vec{r}, \vec{p}; E)$ takes the form

$$\chi(\vec{r}, \vec{p}; E) = e^{i\vec{p}\cdot\vec{r}} + \int d\vec{r} L(\vec{r} - \vec{r}'; E) V(r) \chi(\vec{r}, \vec{p}; E) \tag{5.9}$$

where

$$L(\vec{x}, E) = \int d\vec{t} \frac{1 - 2n_F(t)}{E - \hbar^2 t^2/m} \exp\left[i\vec{x} \cdot \vec{t}\right]. \tag{5.10}$$

In Eq.(5.10) n_F is the Fermi distribution function and m the charge carrier mass. The zero in the denominator has to be handled properly in order to obtain the boundary condition of outgoing waves.

Here we make a further restriction considering χ spherically symmetric. In the language of the scattering theory this corresponds to consider only the s-type contribution to the generalized scattering amplitude $\Gamma(p)$. A simple analytic solution of the integral equation (5.9) can be obtained if we choose a delta function interaction

$$V(r) = \frac{A}{r_0^2}\delta(r - r_0) + \frac{B}{r_1^2}\delta(r - r_1) \tag{5.11}$$

where the first term takes into account the repulsive nature of the interaction at very short distances and the second term represent the short range attractive contribution. These characteristics of the potential reproduce the main features of the polaron-polaron potential. Of course, the delta function form has been chosen for simplicity and more realistic potentials would not change qualitatively any of our results.

Once we have an explicit solution for $\chi(r_0, p)$ and $\chi(r_1, p)$ we can calculate the generalized scattering amplitude (Eq.(5.8))

$$\Gamma(p) = 4\pi[A\chi(r_0, p)j_0(pr_0) + B\chi(r_1, p)j_0(pr_1)]. \tag{5.12}$$

Γ depends on the pair energy $E = (\hbar p)^2/m$ and, through the Fermi distribution function, on the chemical potential μ and temperature T. $j_0(x)$ is the spherical Bessel function of 0 order.

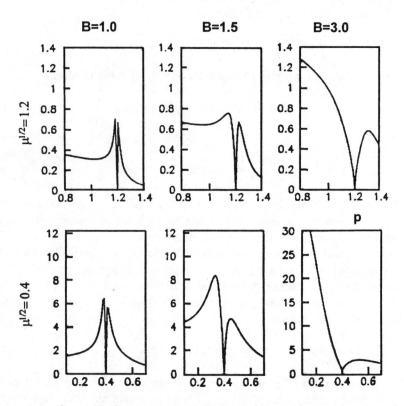

Figure 5.8: Generalized scattering amplitude Γ (see text) as a function of the relative momentum $\bar{p}$. Different columns corresponds to different values of the weight of the attractive part, B, whereas different rows to different values of the chemical potential. Note that the resonance is pinned at $\mu^{1/2}$.

In Fig. 5.8 we plot Γ versus p for a number of values of the chemical potential μ and of the weight of the attractive part B at fixed temperature. The calculation has been performed measuring energies in units $\epsilon_0 = 70 meV$ and lengths in units $R_p = \hbar/\sqrt{2m\epsilon_0} = 4.9A$. The potential used for the calculation is characterized by the following parameters $A = 0.2$, $B = -0.9$, $r_0 = 0.1$ and $r_1 = 1$. The parameters are chosen fitting the asymptotic effective potential (5.5).

Some points have to be noted:

a) a peak in the effective two-particle interaction, Γ is also present at energies lower than μ. They are usually related to the normal state instability of the Fermi gas (Cooper pair problem) and are relevant in BCS-type theories. However the excitations we are interested in are those above μ.

b) A peak at energies slightly larger than μ is always found.

c) The peak narrows as μ increases.

d) The maximum of the curve is pinned on the chemical potential.

This simple analysis suggests the existence of a pair state above the chemical potential μ with a finite life-time which increases with μ. The conclusion is very important since in principle it allows the use of the asymptotic effective potential to calculate ground state properties of the many-polaron system. In particular we are mainly interested in the possibility of resonant bipolaron formation. The previous result will be used in Ch. 7.

Chapter 6

Results of the microscopical model

This chapter presents the numerical results on polaron and bipolaron properties, which can be obtained using the variational procedure introduced in Ch. 4.

6.1 Numerical procedure

The actual calculation of the BPP binding energies and effective masses is organized in several steps. The first step is the self-consistent determination of σ_1 and σ_2 for chosen values of the variational parameters γ and β. The second step consists in the calculation of the total energy, Eq. (4.7), which corresponds to a given $\vec{Q}$. This procedure is iterated for several γs and βs until a minimum in the total energy is found.

It can be shown analytically that for $\gamma = 0$ and $\beta = 0$ the total energy $E_T(Q)$ reduces to that of two free PPs, $E_{as}^{(2)}(Q)$. Therefore it is correct to define the BPP binding energy E_b as the difference between the total energy and the energy of two distinct PPs both calculated at $Q = 0$. Fig. 5.2 exemplifies the statement.

When $E_b \leq 0$ the BPP can form as a bound state. Physically this is due to the competition between the polarization energy and the Coulomb repulsion; in particular the former overwhelms the latter because of the interference between the polarization clouds of the two PPs. The choice of the envelope function (Eq. (4.5)) reflects this competition.

In the limit $Q \to 0$, $E_T(Q)$ and $E_{as}(Q)$ can be written in the form

$$E_T(Q) = E_{T,0} + \frac{\hbar^2}{2M^*}Q^2 \tag{6.1}$$

$$E_{as}^{(2)}(Q) = E_{as,0} + \frac{\hbar^2}{4m^*}Q^2 . \tag{6.2}$$

In the previous equation M^* and m^* are the BPP and PP effective mass respectively. Their values are obtained fitting Eqs.(6.1) and (6.2) to the numerical data. It is found that M^* and m^* are independent on Q over a wide range of the Brillouin zone.

Assuming the plasma frequency not dispersive, we have calculated:

- the total energy $E_T(Q)$ of the biplasmapolaron (BPP);

- the BPP and PP effective masses M^* and m^*, respectively;

- the radius R_b of the BPP.

We use $\hbar\omega_l$ as the unit for energies, R_p as the unit for length, and the band mass m as the unit for the effective masses M^* and m^* (see Sec. 3.4). In these units the above quantities are functions of λ, α and η. We recall that the first parameter is connected to the electronic density through the plasma frequency; its value depends on the electronic density, but also on the band mass of the electron, on the energy $\hbar\omega_l$ of the longitudinal optical phonon and on ϵ_∞. Although the results of this chapter are independent of the values of these quantities, we can fix them to have an easier comparison with the experimental data. From experiments on high-T_cs one can extract the following values: m between 2 and 5 m_e, where m_e is the free electron mass (Sec. 2.5.2), $\hbar\omega_l = 70meV$ and $\epsilon_\infty = 4$ (Sec. 2.5.3). The optical phonons we have in mind are the high frequency ones involving the apex oxygen or the two O_p oxygens. With the previous numbers $\lambda = 1$ corresponds to an electronic density of $n = 4.4\,10^{19}cm^{-3}$. Moreover it is found that η is of the order of some hundredth and finally $\alpha \simeq 8$. This value of α is also supported by the experiments of Mihailovic et al. [Mihail90].

6.2   The BPP binding energy

We minimize the total energy (4.7) with respect to β and γ, using for $f_\mathbf{k}$ and $g_\mathbf{k}$ the solution of the differential equation (4.12) together with the expansions (4.17) and (4.18). Then the BPP binding energy $E_b = E_{T,0} - E^{(2)}_{as,0}$ is calculated for fixed values of α and η as a function of λ. The result is reported in Fig. 6.2. We see that, for $\eta = 0.01$, the binding energy E_b is negative for small λ (the BPP is bound); E_b ends up with positive values as far as λ is increased. It can be concluded that when $\lambda \leq 1$ ($\omega_l \simeq \omega_p/\sqrt{\epsilon_\infty}$), the dynamical cooperative effect between phonons and plasmons increase the stability of the BPP; at larger values of λ the screening effects become more important and E_b vanishes. In correspondence to $E_b > 0$ we find a relative minimum for $E_{T,0}$; it can be argued that the BPP exists as a resonant state.

Concerning the dependence of the binding energy on the electron-phonon coupling constant α, the expected result is displayed in Fig. 6.2; larger α values correspond to larger binding energies.

The effects of the plasmons in the bipolaron problem is clear when the previous results are compared with the analysis performed by Bassani *et al.* [Bassani91]. They found that the bipolaron exists as a bound state when $\alpha \geq \alpha_c$, α_c being 6 in three dimensions and 4 in two; however this condition is not sufficient because η must be smaller than an α-dependent η_c. For $\alpha = 8$, $\eta = 0.01$ and 0.05 are respectively smaller and larger than the η_c calculated by Bassani *et al.*. Therefore it can be concluded that the condition on the values of η is no longer so critical, if the screening effects are taken into account.

As far as the PP self-energy is concerned, Fig. 5.2 shows that the PP self-energy increases with incresing η. Since a larger η implies a smaller ionicity of the lattice, the increase of the self energy is to be ascribes to plasmon exchange. A similar behaviour can be seen also in Fig. 6.2: at small λ values E_b is larger in the case $\eta = 0.05$ with respect to the curve with $\eta = 0.01$; the BPP starts as a metastable state and then becomes bound. Increasing λ, E_b becomes comparable in the two cases as long as the states are bound. We have carried out the calculations as long as $\lambda \leq 2$ because otherwise the dilution parameter nR_b^3 becomes larger than 1, as we are going to see later (R_b is the average bipolaron radius).

6.3   BPP radius

The BPP radius is given by the average

$$R_b = \int r^3 \phi(r)^2 dr \ . \tag{6.3}$$

We find that the BPP radii are nearly constant for given α and η and depend on λ. The dilution parameter nR_b^3 ranges from 10^{-5} for $\lambda = 0.01$ to 1 for $\lambda = 2$ and there are not significant variations with η. From Tabs. 6.1- 6.5 one can see that, as long as $\lambda < 2$, the BPP radius and the dilution parameter are sufficiently small to justify the single pair picture. It can be also seen that the Pippard coherence length $\xi_0 = \frac{\sqrt{3}}{2\pi} R_b$ is small (tens of Å) in agreement with experimental data discussed in Sec. 2.4.7.

The quantities R_b, $k_F\xi_0$ and nR_b^3 as a function of λ are reported in Tabs. 6.1, 6.2, 6.3, 6.4, 6.5 at different α and η values. As a guide to the reading we remind that the actual values of the polaron radius in high-T_c materials is of the order of a few Å.

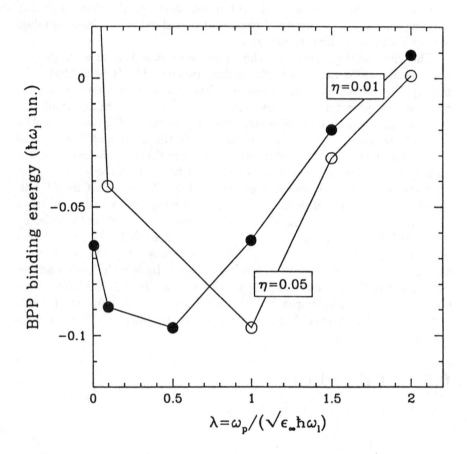

Figure 6.1: The binding energies of the biplasmapolaron are shown as function of λ for $\alpha = 8$ and $\eta = 0.01$, 0.05. $\lambda = 1$ indicates the electronic density of 4.4 $10^{19}\,cm^{-3}$. The energies are in units $\hbar\omega_l$

λ	R_b	$k_F\xi_0$	nR_b^3
0.01	4.8	0.03	10^{-5}
0.1	4.8	0.13	10^{-4}
0.5	4.8	0.38	10^{-3}
1.0	4.8	0.61	0.3
1.5	4.8	0.80	0.8
2.0	4.8	1.00	0.8

Table 6.1: Values of the mean BPP radius R_b in polaronic units, of $k_F\xi_0$, of nR_b^3 as function of λ for $\alpha = 6$ and $\eta = 0.01$.

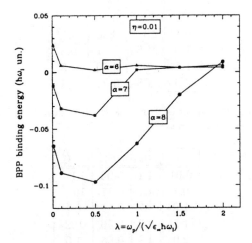

Figure 6.2: The binding energies of the biplasmapolaron are shown as function of λ for $\alpha = 6, 7, 8$ and $\eta = 0.01$. $\lambda = 1$ indicates the electronic density of 4.45 $10^{19} cm^{-3}$. The energies are in units $\hbar\omega_l$.

λ	R_b	$k_F\xi_0$	nR_b^3
0.01	4.1	0.02	10^{-5}
0.1	4.1	0.11	10^{-3}
1.0	3.7	0.45	0.1
1.5	3.6	0.57	0.3
2.0	4.8	0.92	1.2

Table 6.2: Values of the mean BPP radius R_b in polaronic units, of $k_F\xi_0$, of nR_b^3 as function of λ for $\alpha = 7$ and $\eta = 0.01$.

λ	R_b	$k_F\xi_0$	nR_b^3
0.01	4.1	0.03	10^{-5}
0.1	3.7	0.08	10^{-3}
0.5	3.7	0.27	$^{-2}$
1.0	3.7	0.41	0.1
1.5	4.8	0.72	0.6
2.0	4.8	0.88	1.1

Table 6.3: Values of the mean BPP radius R_b in polaronic units, of $k_F\xi_0$, of nR_b^3 as function of λ for $\alpha = 8$ and $\eta = 0.01$.

λ	R_b	$k_F\xi_0$	nR_b^3
0.01	13.2	0.08	10^{-4}
0.1	13.2	0.36	10^{-2}
1.0	4.8	0.61	0.3
1.5	4.8	0.79	0.8
2.0	4.8	0.95	1.4

Table 6.4: Values of the mean BPP radius R_b in polaronic units, of $k_F\xi_0$, of nR_b^3 as function of λ for $\alpha = 6$ and $\eta = 0.05$.

λ	R_b	$k_F\xi_0$	nR_b^3
0.01	4.1	0.03	10^{-5}
0.1	4.1	0.11	10^{-3}
1.0	3.7	0.41	0.1
1.5	3.3	0.50	0.2
2.0	4.8	0.85	1.0

Table 6.5: Values of the mean BPP radius R_b in polaronic units, of $k_F\xi_0$, of nR_b^3 as function of λ for $\alpha = 8$ and $\eta = 0.05$.

6.4   Effective masses

The calculation of the bipolaron effective mass proceeds as already discussed in Sec. 6.1, namely fitting Eq.(6.1) to the numerical data. As far as the plasmon branch is assumed not dispersive, the PP effective mass m^* can be calculated analitically

$$\frac{m^*}{m} = 1 + \frac{\alpha}{6}\left(\frac{F_1^2}{R_1^{\frac{3}{2}}} + \frac{F_2^2}{R_2^{\frac{3}{2}}}\right). \tag{6.4}$$

The limiting behaviour of the previous quantity as $\lambda \to 0$ yields the polaron effective mass in the intermediate coupling regime, $m^* = m(1 + \alpha/6)$.

Fig. 6.4 displays m^* (upper panel) and M^* (lower panel) as a function of λ. First, it has to be noted that m^* and M^* take large values in the low density range. This is due to the fact that the (B)PPs tend to be self-trapped when the screening is not effective in delocalizing the particles. Second, the plots are limited to the range $\lambda > 0.1$. In fact the branch Ω_2 vanishes in the limit $\lambda \to 0$. It is well known that whenever the boson frequency vanishes an unphysical divergency of the mass is introduced, the divergency being removed by taking into account two or more boson processes. As far as our calculation is concerned, we do not attempt to account for those higher order processes. As a consequence the results concerning the PP and BPP cannot be trusted for $\lambda < 0.1$.

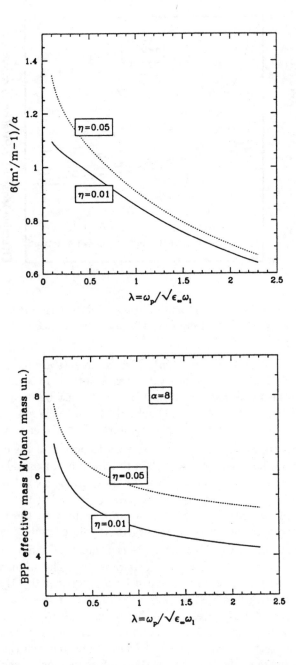

Figure 6.3: Upper panel: the quantity $6(m^* - 1)/\alpha$ is shown as function of λ for $\eta = 0.01$ and 0.05. Lower panel: the biplasmapolaron effective masses M^* are shown as function of λ for $\alpha = 8$ and $\eta = 0.01$ and $\eta = 0.05$.

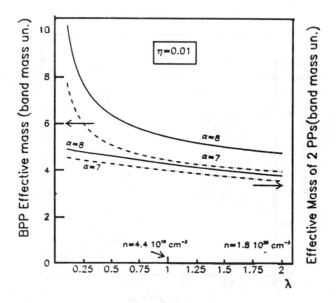

Figure 6.4: The BPP mass M^* and two times the PP mass m^* as a function of λ and for different values of α. Note that $M^* \to 2m^*$ as $\lambda \to \infty$.

Fig. 6.4 shows both M^* and $2m^*$. Two comments are in order. First, M^* tends to $2m^*$ in the limit of large λ values, as expected. In turn, Eq. 6.4 implies m^* tends to the band mass m as $\lambda \to \infty$, as it must be because of the total screening of the electron-phonon coupling. Second, the effective masses are larger the stronger the electron-phonon interaction (higher values of α).

As a final remark on this section, we wish to note that the calculated BPP effective mass is large enough to explain the order of magnitude of the critical temperature in high-T_c materials when the Bose-Einstein condensation formula is used (the typical densities in high-T_c being 10^{20}-10^{21} cm^{-3}). On the other hand this BPP mass values are not so large to prevent the mobility of the biplasmapolarons, which is of course crucial for superconductivity.

The behaviour of the bipolaron effective mass as a function of the carrier density can be compared with experiments. The in-plane and out-of-plane superconducting carrier mass can be measured indirectly from London penetration depth associated with a contemporary measure of the Hall number for different doping levels. This has been done by Athanassopolou and Cooper [Athan95]. Fig. 6.5 reports their experimental values for the easy mass together with the calculated BPP mass. Since our model is isotropic, the geometric average of the in- plane and out-of-plane mass is shown for a more significant comparison. In the comparison it has been assumed that the Hall number is the carrier density. Even though the measured quantities decrease faster than the calculated one, however it is remarkable that the right trend is

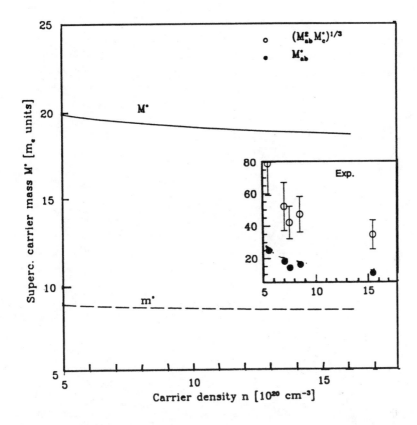

Figure 6.5: Calculated BPP effective mass M^* (solid line) and PP mass m^* (dashed line) for $\alpha = 8$, $\eta = 0.01$. The inset shows the experimental in-plane mass M^*_{ab} as a function of the Hall number (dots) [Athan95]; the geometric average $(M^{*2}_{ab}M^*_c)^{1/3}$ of the in-plane and out-of-plane mass is also shown for comparison.

catched. It must be stressed that no adjustable parameters have been used; in particular a band mass of 2 electron masses has been used in the plot of the calculated PP and BPP mass.

Chapter 7

The Boson-Fermion Model for high-T_cSC: Thermodynamics

We have studied the ground state properties of an almost dilute ($r_s \simeq 1$) system composed by electrons and optical phonons in interaction with each other. The existence of mobile pairs with a short average radius has been discussed in a low-to-intermediate range of the carrier density. The formation of pair resonant states in the presence of a short ranged attractive potential has been shown at higher densities. We wish now to use the concepts described in the previous chapters to study some thermodynamical properties of the system.

First of all, it is useful to summarize the main ingredients to be used in an attempt to rationalize the phenomenon of high-temperature superconductivity:

- a single bipolaron can form for coupling constants $\alpha > 6$ and $\eta \ll 1$ (Ch. 6);

- both static and dynamic screening effects due to the electron gas reduce the bipolaron binding energy (Ch. 6); they reduce also the polaron and bipolaron effective masses, which the bare band mass values in the limit of very high densities (Ch. 6);

- for bound bipolarons, the effective polaron-polaron potential is self-consistently dependent on the parameters of the pair envelope wave-function and has an attractive part (Ch. 5);

- when the bipolaron is unbound, the effective potential reduces to the asymptotic potential described in Sec. 5.4. It is short ranged and it is still characterized by the existence of an attractive well. Such a model effective potential has been shown to have pair resonant states in the presence of the many-particle system (Sec. 5.6, Ref. [Cataud96]).

In the following we assume the existence of well defined resonant states of the bipolarons and we proceed further analyzing the thermodynamical consequences of such hypothesis. The existence of resonant states suggest a model in which polarons and bipolarons coexist, the bipolarons being allowed to decay into two polarons. We call this kind of picture "boson-fermion model" (BFM). It was introduced first by J. Ranninger and S. Robaszkiewicz [RannRobasz85] in the framework of the small polaron scheme and then adopted in a phenomenological approach by R. Friedberg and T.D. Lee [FriedLee].

We wish to apply the Friedberg and Lee formalism, using as ingredients the density dependent energy spectra and effective mass for BPP and PP suggested by the plasmon–bipolaron model.

We assume that the binding (resonant) bipolaron energy $E_b(n)$ and the effective polaron and bipolaron masses $m^*(n)$ and $M^*(n)$ respectively depend on the electron density n as displayed in Fig. 7.1.

This density dependence is consistent with all previous results at low densities (Ch. 6) and is extrapolated at high densities. We anticipate also that, as we have *a posteriori* checked, the detailed density dependence of the above quantities as well as the density value of the crossing between $E_b/2$ and E_F do not change at all the qualitative trend of the results. The only requirement is that the resonant energy be an increasing function of the density eventually crossing E_F.

We note also that the insertion of a Fermi sphere of polarons requires that the envelope bipolaron wavefunction does not contain Fourier components for $k < k_c = \sqrt{2\mu m^*}/\hbar$, where m^* is the polaron mass and μ is the chemical potential at $T = 0$. This phase space reduction gives rise to a further increase of the resonant bipolaron energy, eventually driving it towards the independent polaron Fermi energy $E_F = \hbar^2 k_F^2/2m^*$.

7.1 Getting started: the Hamiltonian

The starting BFM hamiltonian is

$$
\begin{aligned}
H \;=\; & \sum_{\vec{Q}}(E_b(n) + \frac{\hbar^2 Q^2}{2M^*(n)} - 2\mu)b_{\vec{Q}}^{\dagger}b_{\vec{Q}} \\
& + \sum_{\vec{q}}(\frac{\hbar^2 q^2}{2m^*(n)} - \mu)c_{\vec{q}}^{\dagger}c_{\vec{q}} \\
& + \frac{1}{\sqrt{V}}\sum_{\vec{Q},\vec{q}}(g(\vec{Q},\vec{q},n)b_{\vec{Q}}^{\dagger}c_{\frac{\vec{Q}}{2}+\vec{q},\uparrow}c_{\frac{\vec{Q}}{2}-\vec{q},\downarrow} + h.c.) \,.
\end{aligned}
\tag{7.1}
$$

The $b_{\vec{k}}^{(\dagger)}$ and $c_{\vec{k}}^{(\dagger)}$ are the destruction (creation) operators for bosons and fermions, respectively. The quantity $g(\vec{Q},\vec{q},n)$ is responsible for the decay of a bipolaron

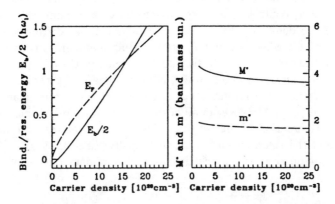

Figure 7.1: Total charge carrier density dependence of half the bipolaron binding(resonant) energy E_b (on the left) and of the bipolaron (M^*) and polaron (m^*) effective masses (on the right). The crossing of $E_b/2$ with the Fermi energy E_F of the free polaron gas is also shown. In this letter we have used a band mass $m = 5m_e$ where m_e is the electron mass.

into a pair of free polarons and vice-versa and depends on the center of mass momentum $\vec{Q}$, on the relative momentum $\vec{q}$ and on the density n. As in [FriedLee] we assume that the polaron and bipolaron chemical potentials are related by the "chemical" equilibrium condition $2\mu = \mu_{BP} = 2\mu_P$.

Following Friedberg and Lee [FriedLee] we can diagonalize the Hamiltonian (7.1) in an approximate way to obtain renormalized free bipolarons and free polarons which are coupled via the macroscopic occupation number $|B|^2$ of the $Q = 0$ bipolaron state. The renormalized polaronic spectrum is

$$E(q) = [(\frac{\hbar^2 q^2}{2m^*} - \mu)^2 + (g(0,q,n) \mid B \mid)^2]^{\frac{1}{2}} , \qquad (7.2)$$

which has the same structure as that of the excitation spectrum in BCS theory: here the squared gap energy Δ^2 is given by $\Delta^2 = g^2|B|^2$. It is important to stress that B drives both the condensate fraction and the superconducting energy gap.

Within the scheme described previously, $g(\vec{Q}, \vec{q}, n)$ can be estimated from the actual shape of the polaron-polaron effective potential. Moreover, consistently with our preliminary calculations, we assume for $g(0, q, n)$ the specific form

$$g(0, q, n) = g_0(n)e^{-\frac{(q-k_c)^2}{k_i^2}} . \qquad (7.3)$$

Eq.(7.3) says that the highest amplitude for the decay of the bipolaron occurs
when the outgoing polarons have an energy pinned at the chemical potential at $T = 0$. The density dependence of the effective potential is contained
in g_0, whereas k_t controls the energy extension of the bipolaron-polaron interaction. An order of magnitude for k_t can be estimated from the relation
$k_B T_c = \hbar^2 k_t^2 / 2m^*$, where k_B and T_c are the Boltzman constant and the critical temperature respectively. In fact, at lower energy the decay is forbidden
by the Pauli exclusion principle while at higher energy by the fact that the
Fourier transform of the repulsive tail of the effective polaron-polaron potential rapidly vanishes. Moreover in all of the calculations we assume that $g_0(n)$
is a decreasing function of the density. This is consistent with the increasing
dependence of the resonant energy on the density; in fact the two quantities
are related in the equation which defines the imaginary part γ of the BPP
state

$$\gamma = \frac{g^2(n) m^{2/3}}{\pi \hbar^3} \sqrt{\frac{E_b(n)}{2}} \,, \tag{7.4}$$

as derived by Friedberg and Lee [FriedLee]. Anyways, the specific dependence
of $g_0(n)$ on the density is less important. The consequences of a constant or
increasing function choice for $g_0(n)$ are discussed later.

7.2   Equations for μ and $|B|^2$

In the scheme of the gran–canonical ensemble it is possible to determine the
macroscopic occupation number $|B|^2$ and the chemical potential μ requiring
that the free energy is minimized and the total number of electrons is conserved. One obtains

$$\mu = \frac{E_b}{2} - \frac{1}{4V} \sum_{\bar{q}} \frac{g(0, q, n)^2}{E(q)} \tanh \frac{\beta E(q)}{2} \tag{7.5}$$

$$n = 2| B |^2 + \frac{2}{V} \sum_{\bar{q}} \frac{1}{e^{\beta(E_b + \frac{\hbar^2 q^2}{2M^*} - 2\mu)} - 1} \tag{7.6}$$

$$+ \frac{1}{V} \sum_{\bar{q}} \frac{1}{E(q)(1 + e^{-\beta E(q)})}$$

$$[E(q) + \mu - \frac{\hbar^2 q^2}{2m^*} + (E(q) - \mu + \frac{\hbar^2 q^2}{2m^*}) e^{-\beta E(q)}] \,,$$

where $\beta = 1/k_B T$. It is worth noting that the quantity $|B|^2$ counts the number
of condensed bosons and at the same time drives the order parameter $\Delta = g|B|$.

Eqs. (7.5) and (7.6) have been studied in some detail by Friedberg and
Lee assuming E_b, m^*, M^* and $g(0,q)$ to be density independent parameters.

We now extend their analysis to take into account the density dependence described in the previous chapters.

7.2.1 A comment on the self-consistent solution

At this point a comment should be made. The calculation of thermodynamic quantities involves the knowledge of the chemical potential and of the resonant energy; on the other hand the bipolaron resonant energy depends on the value of the chemical potential itself. It follows that the problem must be treated self-consistently at any given temperature. In principle we should solve self-consistently Eqs. (7.5) and (7.6) for μ and n together with Eqs. (5.8) and (5.10), which give the resonant BPP energy for fixed μ and also with Eq. (7.4) for the width of the resonant state. Since the resonance is not sufficiently sharp, it is difficult to achieve the full self-consistency. However it is worthwhile to note that, if this is the case, the resonant energy has been found to be pinned at the chemical potential (at $T = 0$); since μ is an increasing function of the density n, this strongly indicate that E_b is an increasing function of n.

7.2.2 The physical picture

The physical picture we have in mind is schematically depicted in Fig. 7.2. We have already shown in Fig. 5.2 the meaning of bound and resonant states at $T = 0$. When a bound state exists, our picture predicts the chemical potential to be at the level of the bound state. As soon as the bound state (in the sense discussed in the previous chapter) disappears, we may have resonant bipolaron states as described in Sec. 5.6. Thus we consider that our system is composed of coexisting polarons and resonant bipolarons.

At $T > T_c$ we have normal polarons which fill energy states up to the chemical potential μ according to the ideal Fermi distribution (Fig. 7.2). We also have (*non condensed*) bipolarons at an energy E_b which must be larger than 2μ; the parabola in the figure depicts the bipolaron band, its curvature being connected to the bipolaron effective mass. The larger is μ, the smaller will be the occupation of the bipolaron state.

As far as the temperature is lowered below T_c, a gap develops in the system and we have the following picture. Polarons fill energy states up to the value of the chemical potential, which lies below the many-polaron normal-state energy (indicated with μ_n). At the same time we have a bipolaron band nearly resonant with the polaron state ($E_b/2$ being larger than or equal to μ); the $Q = 0$ bipolaron state is macroscopically occupied. For this configuration to be the ground state, it must happen that the total energy of the polaron and bipolaron state, weighted with the appropriate statistical distribution, must be smaller than the energy of the normal state μ_n.

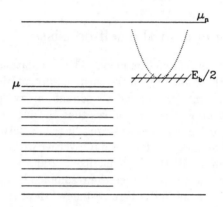

Figure 7.2: Schematic picture of the BFM for plasmapolarons and biplasmapolarons. PPs fill fermionic energy levels up to μ. BPPs occupy the level at $E_b/2 > \mu$, the level having a given width. The parabola indicates non-condensed bosons and μ_n is the chemical potential of the normal state.

As far as the density increases, $E_b/2$ and μ both shift towards μ_n remaining always almost sticked to each other and the bipolaron state empties in favour of the polaron subsystem. The limiting cases are when $E_b/2$ crosses μ_n and thus the system becomes normal. On the other hand, when $E_b/2 < 0$ μ must be negative and only condensed bipolarons exist.

7.3 Comparison with experiments

7.3.1 Critical temperature as a function of the total density

Let us now focus our attention on the critical temperature T_c as a function of the total electron density n. T_c is calculated solving Eqs. (7.5–7.6) with $|B| = 0$. This is displayed in Fig. 7.3. Different curves correspond to different linear dependences of g_0 on λ (and then on the square root of the density of holes per CuO_2 unit, x). The critical temperature vs. the total electron density (in the figure it is displayed the total number of holes per CuO_2 unit) clearly reproduces the typical bell shape found in high-T_c superconductors: the critical temperature first increases, then reaches a maximum, and eventually decreases to a vanishing value. In the present model the vanishing of the critical temperature for high densities is due to the crossing of $E_b/2$ with

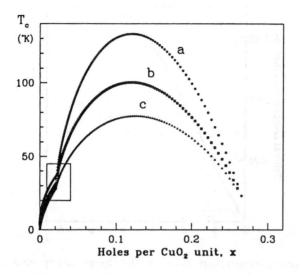

Figure 7.3: Critical temperature T_c vs. x, the charge carrier number per CuO_2 unit, for three different density dependent g_0 (see text): (a) $g_0 = 10 - 19.3x^{1/2}$; (b) $g_0 = 5 - 9.6x^{1/2}$; (c) $g_0 = 2 - 3.7x^{1/2}$, g_0 being measured in units of $\hbar\omega_l$ (we have used $\hbar\omega_l = 70$ meV).

E_F [Blatt64]. In the low density regime, when the bipolaron binding energy is larger, an electron density increase corresponds to a larger occupation of the bipolaron $Q = 0$ state ($|B|^2$); this produces higher critical temperatures. In this regime there are very few polarons in equilibrium with many bipolarons and the critical temperature follows the expected behaviour for an ideal boson gas.

Increasing the density in the overdoped regime, a smaller number of bipolarons occupies the $Q = 0$ state and the transition temperature decreases.

A final comment on Fig. 7.3 is devoted to the marked box, which seems to contain a discontinuity in the derivative dT_c/dn. This discontinuity can be verified analitically simply performing the derivative with respect to n of both members in Eq.(7.5). After some lenghty algebra it can be verified that the discontinuity depends on k_c which is contained in $g(q)$. In fact k_c is zero as long as the chemical potential is negative, while it is $k_c = \sqrt{m^*\mu}$ when μ is positive. The density value at which the discontinuity appears is indeed the same of the crossing of μ through zero. The discontinuity is therefore an artifact of the model for the $g(q)$ function. Fig. 7.4 shows for instance the

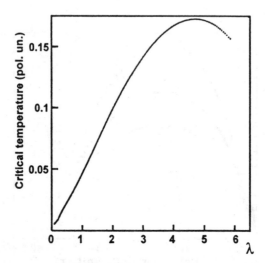

Figure 7.4: Critical temperature T_c vs. x, the charge carrier number per CuO_2 unit, for $g_0 = 10 - 19.3x^{1/2}$ in the case in which μ is always positive (see text).

critical temperature calculated making the BPP state resonant in the whole range of densities; in this case the resonant energy crosses two times the Fermi energy of the independent polaron gas.

The previous results are encouraging. The parameters entering Eqs.(7.5) and (7.6) are microscopically calculated and then extrapolated at higher densities (the detailed dependence being anyways unimportant). However the values of $g_0(n)$ are difficult to extract from experiments. This fact introduces some degree of arbitrariness; for instance, the maximum critical temperature T_c^{max} depends on these quantities. Nevertheless, the dependence of T_c vs. n is a basic feature since it hinges only on the existence of a crossing between E_b and E_F. In order to reduce as much as possible the spurious effects arising from the arbitrariness of the chosen parameters, we display in Fig. 7.5 the ratio T_c/T_c^{max} as a function of the total electron density for different $g_0(n)$ and compare the theoretical curves with experimental data for a number of high T_c materials [ZhSato93]. It can be seen that the agreement is quite satisfactory; moreover the calculated ratios show a remarkable weak dependence on $g_0(n)$ (see the main figure) and k_t (see the inset). From the comparison between Fig.7.3 and Fig.7.5 it can be seen that in Fig.7.5 the calculated curves have been rigidly shifted to the right by the amount $\Delta x = 0.06$ (x being the number of charge carriers per CuO_2 unit which is proportional to n). This is required to reproduce the experimental threshold which can be attributed to localization by crystal disorder.

Concluding, Fig. 7.5 displays the comparison of the theoretical T_c curve

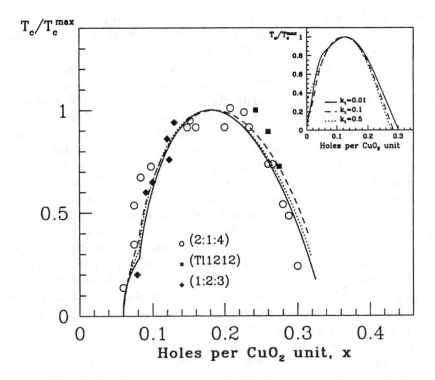

Figure 7.5: Normalized critical temperature T_c/T_c^{max} vs. charge carrier number per CuO_2 unit, x, for the three different g_0 indicated in fig.7.3 (solid line: (a); dotted: (b); dashed: (c)). The theoretical curves are compared with the experimental data of Zhang and Sato [ZhSato93] for different high-T_c compounds. Inset: the same quantity as in the main figure for different k_t values (see text) keeping $g_0 = 2 - 3.71x^{1/2}$, g_0 being measured in units of $\hbar\omega_l$.

with the experimental data for a number of cuprates, which show a characteristic universal behaviour. Fig. 7.5 contains two fitting parameters: the onset edge of T_c and the density which corresponds to the crossing between the bipolaron resonant energy E_b and the polaron $T = 0$ Fermi energy E_F.

7.3.2   Critical temperature as a function of $|B|^2/M^*$

Another peculiar experiment by Uemura *et al.* and Niedermayer *et al.* [Uemura93,Nieder93] concerns the magnetic penetration depth λ of high-T_c materials in the overdoped region. The experiment has been already discussed in Sec.2.4.1, where the characteristic loop shape of T_c as a function of $1/\lambda^2$ is displayed. In the clean limit λ^{-2} is proportional to the ratio n_c/m_0 of the superfluid density to the charge carrier effective mass. In the case of anisotropic materials like the cuprate family, m_0 is the easy in-plane mass. The experimental data show an about universal linear behaviour in the underdoped region (low total electron density), while at optimal doping the curve turns back to zero through lower critical temperatures. Our model is able to reproduce this behaviour at least qualitatively. In Fig. 7.6 we show T_c vs $|B|^2/M^*$.

It must be noted that the calculated slope in the underdoped region is significantly larger than the measured one (a factor of about three). This discrepancy is actually due to the fact that our model does not take into account the mass anisotropy while the experiments definitely involve the in-plane mass (at least for the cuprate family). Our conclusion is also supported by the comparison between Fig. 7.5 and Fig. 7.6: in the former any mass is not directly involved in the theoretical curve while the opposite happens in Fig. 7.6. Moreover, if one puts some numbers in the ratio $M_{\parallel}/M_{\perp}$ for the high-T_c (which is of the order of 10^{-3} for the overdoped materials), it is easily seen that the factor of three in the slope is restored.

We stress that this result is qualitatively reproduced if we modify the density dependence of the decreasing function $g_0(n)$. In particular if we use a constant g_0, the underdoped branch of the curve is nearly overlaped by the overdoped branch; if $g_0(n)$ is an increasing function of the density the overdoped branch is above the underdoped one.

Therefore, one can conclude that, as far as these critical properties are concerned, the anisotropy does not play any relevant role and the discrepancy can be absorbed into a factor in the carrier mass due to the anisotropy. We find that the calculated slope depends on $g_0(n)$ and k_t and reaches the lowest value for $g_0(n) \to 0$ and $k_t \to 0$. In this limit the slope shows a strong insensitivity on all of the other parameters and can be expressed as

$$\tilde{T}_c \simeq a \frac{|B|^2}{M^*} ,$$
(7.7)

a being $\simeq 23$. In Eq.(7.7) we have used polaron units and therefore T_c is measured in unit $\hbar \omega_l/k_B$ and $|B|^2$ (which is proportional to the superfluid

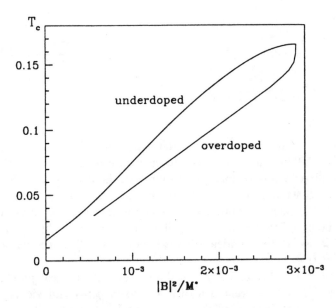

Figure 7.6: $\tilde{T}_c = T_c/\hbar\omega_l$ vs. the ratio of the bipolaron condensate to the bipolaron effective mass in polaron units (see text). Here, as in curve (c) in fig.7.3, $g_0 = 2 - 3.7x^{1/2}$ is assumed.

density [FriedLee89b]) in units of R_p^{-3}; furthermore M^* is the renormalized bipolaron mass (in units of band mass).

7.3.3   Study on $\mu(T, n)$

The behaviour of μ at the critical temperature is an important test for the model. It is known from the experimental work by D. van der Marel and coworkers [Marel92] that there is an universal jump in the slope of the chemical potential at the transition temperature. An interesting point is that, in the present model, the jump exists and is a direct consequence of the discontinuity in the derivative $\partial|B|^2/\partial\beta$ at the critical temperature $1/\beta_c$.

The proof can be derived just from Eq.(7.6), bearing in mind that Eq.(7.5) becomes meaningless for $T > T_c$. We carry out the derivative of both members in Eq.(7.6) with respect to β for $\beta \to \beta_c^+$ and $\beta \to \beta_c^-$ and then we set $\beta = \beta_c$. After a lenghty algebra one obtaines the following result

$$\frac{1}{k_B T_c}\left[\frac{\partial(\mu_n - \mu_s)}{\partial\beta}\right]_{\beta=\beta_c} = \left[\frac{2\pi^2 + \frac{g_0^2(n)}{2}\int_0^\infty \Phi(q)dq}{\frac{m^{*\,3/2}}{\beta_c}\int_0^\infty \Xi(k)dk}\right]\left[-\frac{\partial|B|^2}{\partial\beta}\right]_{\beta=\beta_c^-} \qquad .(7.8)$$

The quantities μ_s and μ_n are the chemical potentials of the superconducting and of the normal state respectively; μ_c is the chemical potential at the critical temperature and:

$$\Phi(q) = \pm q^2 g(0, q, n) \frac{(1 - e^{-\beta_c|\omega_q - \mu_c|})^2 - 2\beta_c e^{-\beta_c|\omega_q - \mu_c|}(\omega_q - \mu_c)}{2(\omega_q - \mu_c)^2(1 + e^{-\beta_c|\omega_q - \mu_c|})^2}$$

$$\Xi(k) = \frac{k^2 e^{-\beta_c|\omega_k - \mu_c|}}{(1 + e^{-\beta_c|\omega_k - \mu_c|})^2} ,$$

$$(7.9)$$

where $\omega_q = q^2/m^*$ and the sign in the first equation is driven by $sign(\omega_q - \mu_c)$.

From Eq.(7.8) it follows that the jump in the derivative of the chemical potential is directly proportional to the jump in the derivative of the condensate fraction. Since it turns out that $|B|^2$ has a finite jump in slope at T_c one can conclude the chemical potential has a discontinuity in its derivative at the transition point. This result is completely different from the one known in the case of Bose-Einstein condensation, where μ has a discontinuity only in its second derivative.

We show in Fig. 7.7 and Fig. 7.9 the calculated chemical potential as a function of T/T_c for different densities and fixed $g_0(n)$ and viceversa. The calculations are compared with the experimental data by van der Marel. Unfortunately, we do not know the actual total carrier density of the measured samples, even though the authors of Ref. [Marel92] conclude from other considerations to be near optimum doping. If this is the case, the agreement would be satisfactory. This is evident from the optimum doping density which appears in Fig. 7.8, where the correspondent critical temperature as a function of n is displayed. As pointed out by Rietveld and Van der Marel [Rietveld92], it is worthwhile to note that this jump in the derivative of the chemical potential with respect to the temperature is related to the jump in the specific heat by the very general relation:

$$\frac{1}{c_n - c_s} \frac{d(\mu_n - \mu_s)}{dT}\Big|_{T=T_c} = \frac{dlnT_c}{dn} .$$

$$(7.10)$$

7.4   Study on $|B|^2(T, n)$

In this section we present some results on the condensate fraction as a function of the temperature for different densities values. They are shown in Fig. 7.10. At fixed density, the condensate fraction decreases as T approaches T_c. For an easier comparison, Fig. 7.11 displays the condensate fraction normalized to its $T = 0$ value as a function of the reduced temperature T/T_c. The correspondent measured quantity [Uemura93] at different doping levels is also shown by the symbols.

From the plot it is also evident that a larger the density leads to a smaller fraction of condensed bosons. This ia due to the shorter BPP lifetime with

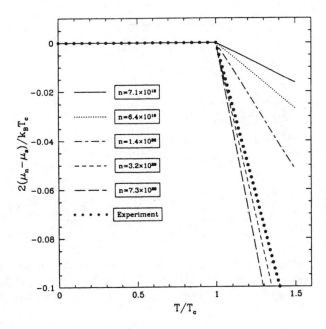

Figure 7.7: Difference in the chemical potential between the normal and the superconducting state $\mu_n - \mu_s$ normalized to $k_B T_c$ as a function of T/T_c for different densities. $g_0(n)$ as in curve (a) of fig.7.3 is assumed. The points represent the experiment from Ref. [Marel92]

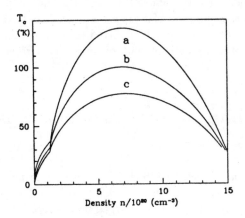

Figure 7.8: The critical temperature vs. n which corresponds to Fig. 7.7. Note the value of the optimum doping density.

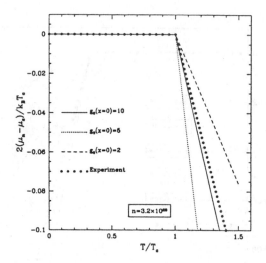

Figure 7.9: Difference in the chemical potential between the normal and the superconducting state $\mu_n - \mu_s$ normalized to $k_B T_c$ as a function of T/T_c for different $g_0(n)$ and fixed density. The points represent the experiment from Ref. [Marel92]

increasing the density. A final comment to this section concerns dependence of $|B|^2(T/T_c)$ on the different $g_0(n)$ considered. From Fig.7.12 it can be seen that this dependence is very slight.

Another significant plot (Fig. 7.13) shows three characteristic quantities: the condensate fraction, the chemical potential μ and the fraction of excited biplasmapolarons as a function of T/T_c for a given density. It is clear that the fraction of "pseudo-fermions" can be inferred as the difference (1−Condensate fraction−fraction of excited BPPs) and, of course, it increases when increasing the density.

7.5 Screening properties of polarons and bipolarons

The Boson–Fermion Model described so far implies an unequal treatment of the screening effects due to the bosons with respect to fermions. In particular it has been implicitly assumed that charged bosons and fermions do screen the interactions in the same way. In fact the binding and/or resonant bipolaron energy is taken to be at any temperature a function of the total carrier density (bipolarons+polarons). This corresponds to assume that both kind of carriers show the same properties, from the point of view of the many-body system. However whenever some BPPs are formed the number of PPs participating

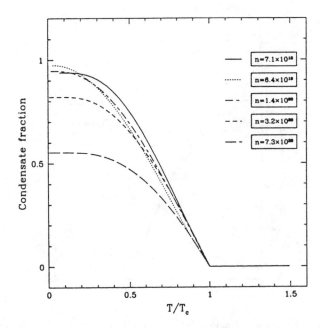

Figure 7.10: Condensate fraction $|B|^2/n$ as a function of T/T_c for different density values. Here the $g_0(n)$ dependence of curve (a) in fig.7.3 is assumed (see text).

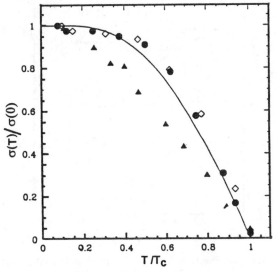

Figure 7.11: Normalized condensate fraction $\sigma(T)/\sigma(0) = |B(T)/B(0)|^2$ as a function of T/T_c for the optimum doping density. Symbols are the normalized experimental $1/\lambda^2(T)$ for Tl2212 as a function of T/T_c at different dopings. The circles refer to the sample near the optimum doping. The experiment is after Ref. [Uemura93]. The Hall effect measurements on the same samples are reported in Ref. [Kubo91].

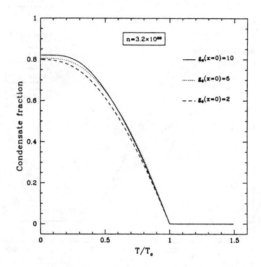

Figure 7.12: Condensate fraction $|B|^2/n$ as a function of T/T_c for the different $g_0(x)$ forms as in fig.7.3. Here the value $n = 3.2 \times 10^{20}$ is assumed.

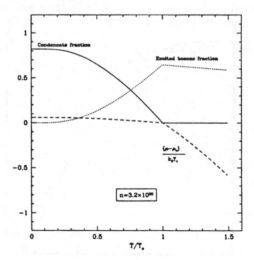

Figure 7.13: Condensate fraction $|B|^2/n$, chemical potential $\mu - \mu_c$ [$\mu_c = \mu(T = T_c)$] normalized to $k_B T_c$, and fraction of excited BPPs as function of T/T_c; $n = 3.2 \times 10^{20}$ is assumed.

to the screening decreases, leading to a reduction of the screening itself. This effect is known, for instance, in the case of the excitons when the screening associated to them [Mahan] is negligible compared to that of the electrons. The previous assumption should be justified.

As far as the present particular case is concerned, it can be shown [Iadonisi96a] that the bipolaron binding energy which results from a boson-driven screening is almost indistinguishable from that which results from the case where only fermions participate. This is true within both the Thomas–Fermi and the Random Phase Approximations for the dielectric function (see Sec. 3.6).

From a more general point of view, one may ask how does a charged boson fluid behaves from the point of view of its screening properties. In particular it is interesting to investigate similarities and differences with respect to the analogous (and well studied) fermion fluid. This could help in disentangling those aspects of the problem which are mainly due to the difference in the quantum statistics.

This problem has been the subject of extensive distinct studies, reported in a number of references [Chiofalo94,Conti94,Chiofalo95,Chiofalo96]. Charged bosons have been investigated using different static approximations for the dielectric function. The main conclusions on the subject are reported in Appendix B.

Chapter 8

Conclusions

8.1 Summary of the main results

The present thesis deals with a study on the formation and properties of pairs of electrons or holes in real space, when they are in a low density system composed of charged particles embedded in a strongly polarizable lattice. In particular it has been shown that the cooperative effect of long range ionic and electronic excitations leads to a *short-range attractive* interaction. Since high-temperature superconductors are characterized by a strong ionicity, a low carrier density and a short coherence length, such a short-ranged attractive interaction is proposed to be responsible for the formation of the superconducting *mobile* pairs.

The study of the system is based on an hamiltonian approach within a two-particle picture. The hamiltonian contains the mutual interactions between the two charged particles, a plasmon and a phonon field. The coupling of the charged particles with the longitudinal optical phonons and with the plasmons are taken to be of the long-range Fröhlich type and the two bosonic fields are treated as coherent states. The hamiltonian can be diagonalized in terms of renormalized plasmon-phonon frequencies and coupling constants. Furthermore the total momentum of the two particles plus the plasmon-phonon field is conserved and therefore it is possible to disentangle the center of mass motion from the relative motion of the two particles. This allows a natural definition of the pair effective mass, which turns out to be of the right order of magnitude needed to explain the high transition temperature with such low densities. The average radius of the pair is found to be of the order of the coherence length in high-T_cSC . As far as the the relative motion is concerned, this is driven by the phonon-plasmon distribution functions, which readjust following the changes in the average distance between the two particles.

The hamiltonian approach has been shown to be equivalent to a formulation based on an appropriate dielectric function in the Random Phase Approximation (RPA). The plasmon-phonon renormalized frequencies are the zeros of the dielectric function and the renormalized coupling constants in the one- particle picture are the residues of $1/\epsilon$.

The basic microscopical parameters which enter this model are the optical phonon frequency, the high and low frequency dielectric constants and the band mass of the undoped material. In addition, the response of the system depends on doping. A variational procedure has been used to calculate the total energy of the pair. The procedure consists in the minimization of the total energy with respect to the phonon-plasmon distribution functions and with respect to an envelope function for the two particles. The first minimization leads to two differential equations which are solved numerically. Since the energy of the two independent particles $E_{as}^{(2)}$ is the total energy in the limit of a uniform pair envelope function, the difference between the total energy and $E_{as}^{(2)}$ gives the pair binding energy. We have called these pairs biplasmapolarons, because they are pairs of electrons plus the polarization effects due to phonon and plasmon interaction. The pair binding energy has been shown to be larger the greater the ionicity or the electron-phonon coupling. Furthermore, it is a non-monotonic function of the density, being larger the closer are the plasmon and phonon frequencies. Correspondingly, the pair as well as the single particle effective mass decreases monotonically with the density.

The two-particle scheme has allowed the determination of an effective potential which depends self-consistently on the pair wave-function. Therefore the study of the system may reduce to the solution of a Schrödinger-like equation with such an effective attractive potential.

The definition of the effective potential allows in principle the study of the system even when the pair binding energy vanishes and it takes positive values. In correspondence of such densities the absolute minimum of the total energy in the space of the relevant parameters becomes a local minimum, which tends to disappear quickly. The bound state is then expected to become a resonant state. The effective potential still depends on the phonon and plasmon distribution functions, even though the dependence on the two-particle envelope function is of course lost. Nevertheless, the effective potential has a *short-range* attractive well, a Coulomb part at small distances and an exponential tail at long distances. The tail is characterized by a screening length which is given by the polaron radius at low density and by the plasmaron radius at large density. Furthermore, it is analytical in the microscopical parameters involved. Along these lines, one of the most important conclusions is that the present work provides an easy-to-use and meaningful physical picture for the microscopical two-particle attractive interaction in a system with long-range phonon and electron degrees of freedom treated on the same footing.

The calculation of the resonances of such a potential in the presence of the many-particle system is a current subject of study. More generally, the properties of a Fermi fluid with an attractive interaction are one of the most discussed and difficult subjects of debate. Considering the existence of the resonances, it is consistent to assume that the energy of the resonant state scales with the Fermi energy of the free plasmapolaron gas. On this basis, the BPP and PP energy spectra and a parametrized effective potential have been used as ingredients in a phenomenological boson-fermion model to calculate thermodynamical properties of such a system .

The model described so far has been applied to the physics of high-temperature superconducting cuprates. The motivation comes from the analysis of the phenomenology of cuprate materials. They basically belong to a family of insulating ionic compounds which become superconducting under heavy doping. A maximum critical temperature can be achieved at an optimum doping level, the correspondent carrier density usually being one or two order of magnitude lower than in typical metals. As far as the microscopical pairing mechanism is concerned, the existence of the isotope effect in cuprate superconductors is undoubtely a fingerprint of the role of phonons. Beyond the isotope effect, a large collection of infrared and Raman spectroscopy data, together with high energy tunnelling measurements and photoinduced absorption measurements relating to insulating compounds points towards the existence of electron-phonon interaction and in particular polaronic effects. In addition the short coherence lengths found in high-T_c superconductors poses the problem of real space pair formation in a Fermi system with a short range attractive interaction. All the previous considerations together with the characteristic dependence of the superconducting properties on the doping level, constitute the basis for the application of the present model to high-T_c superconductivity. In particular, several calculated quantities like the critical tememperature, the chemical potential, the condensate fraction as a function of the carrier density have been found to be in satisfactory agreement with the experimental data.

The screening properties of the charged Bose gas have been the subject of a distinct study using a dielectric function as well as a sum rule approach. The study has revealed that a Random Phase Approximation for the charged Bose gas at zero temperature is accurate only at very small values of the coupling strength r_s because the ideal gas is fully condensed in the zero momentum state. Static local field theories like STLS or VS are more suitable. Comparison with Monte Carlo data shows that the former is very accurate in the description of the short range behaviour of the pair correlation function whereas the latter accounts for the long wavelength behaviour of the static dielectric function, since it satisfies the compressibility sum rule. Both of them are accurate in their predictions on the ground state energy. The main point with the charged Bose fluid is the negative compressibility at all r_s values. This fact has striking consequences on the existence of overscreening effects due to pile-up of the particles, because of the statistics. Other striking consequences are evident in the negative dispersion of the collective excitations at low $k-$ vectors; it is evidently connected to the fact that the ground state energy for bosons is entirely due to the correlation energy. As far as the comparison with the fermion fluid is concerned, major differences in the structure of the fluids appear for small values of the coupling strength r_s while they tend to disappear for values of the coupling strength greater than about five. The static screening charge is characterized by the presence of overscreening effects in the Bose fluid and by Friedel oscillations due to the discontinuity at the Fermi edge in the fermion fluid. The sum rule approach has been shown to be useful in order have a systematic study of infrared divergences in density and particle excitations. This is summarized in Tab. B.1. In particular it is worth to point out that the existence of the condensate leads to the consideration of sum rules for density-density, particle-particle and mixed density-particle excitations.

8.2   Open problems

At this point some comments on open problems raised by our approach are in order. First, the two-particle scheme could be considered in principle too rough. However the same picture (without the phonon interaction) has been applied to the problem of exciton formation and bleaching [Ninno94,Capone94]. The application has been successful in reproducing the results of more sophisticated many-body techniques. The agreement is due to the low density and large band mass involved in the specific problem.

Furthermore, the model has been studied in the case of a cubic lattice. This is quite far from the reality of high-T_c cuprates, which are layered materials. Experiments seem to indicate that no correlation exists between the layered structure and the nature of the microscopical mechanism (Ch. 2). The superconducting properties calculated within the present model do not suffer from the assumed isotropy. However, a more realistic anisotropic model is clearly needed. Some calculations on the anisotropic large plasmapolaron have been already carried out by Iadonisi et al. [Iadonisi96b]. In this case three boson fields are present: an anisotropic plasmon field and two optical phonon fields. The two phonon fields are generated by atoms moving in direction parallel and perpendicular to the CuO_2 planes, respectively. Using the right microscopical parameters (band mass, optical constants and phonon frequency either in or out of the planes), it turns out that the phonons polarized perpendicular to the planes are weakly screened by the electron gas, which is mainly confined in the CuO_2 layers. It is worthwhile to comment that the introduction of the near two dimensional layered structure in the theory is needed to tackle the study of the normal state properties of high-T_c cuprates. For instance Alexandrov and Mott [Mott93] have shown how the puzzling temperature dependence of the resistivity and the Hall effect can be explained within a quasi-2D model of extended (free) bipolarons.

The problem of the symmetry of the superconducting wave-function has not been addressed here. An isotropic s-wave has been assumed. In principle a biplasmapolaron wave-function with higher orbital angular momentum L can be studied. This has been already done by Bassani et al. [Bassani91] in the case of a bipolaron without the inclusion of screening effects. The authors of Ref. [Bassani91] found that bipolaron formation with $L = 2$ is less favoured than the $L = 0$ case.

The large polaron approach deserves a further comment. It has been already noted that typically large polaron radii in high-T_c cuprates are of the order of the lattice constant; this pushes the present approach on the limit of validity. The difficulty can be in principle overcome inserting band structure details in the kinetic energy term of the hamiltonian. However a simple argument has shown that the consideration of screening effects certainly brings the system back in a large polaron picture (Sec 5.3).

Finally, the optical properties of large plasmapolarons with an attractive short range interaction may constitute another subject of study, in order to have a deeper understanding of the optical conductivity experimentally found in high-T_c cuprates [Calvani96,BuchWach]. Optical properties of large polarons and bipolarons have been studied by Devreese et al. [Devreese84,Devreese96] and the characterization of the MIR-band in terms of relaxation of excited bipolaron states has been given. Emin [Emin93] has also concluded that the existence of a Drude peak and a temperature independent d-band in the optical spectra of high-T_c cuprates points towards the relevance of *large* polarons or bipolarons. In fact in the case of large polarons and bipolarons the Drude peak edge is slightly below the typical phonon frequency and moreover the d-band is temperature independent, it being due to transitions to excited polaron states. In the small polaron or bipolaron case, the d-band would be temperature dependent and the Drude edge would be severely pushed towards very small frequencies. Anyways, the large polaron picture of the optical conductivity is expected to be different as far as the effects of the electron gas are taken into account. The treatment of electrons and phonons on the same footing may have striking consequences on the frequency dependence of the electron self-energy. This problem is interesting by its own and deserves some future work.

Appendix A

Relationships between F_i and R_i

Here are listed some of the relations that can be derived among the renormalized frequencies and coupling constants F_i and R_i. First let us define the quantity:

$$T_i = \sqrt{(1 - R_i^2)^2 + \lambda^2(1 - \eta)} \; .$$

Then:

$$R_1 R_2 = \lambda\sqrt{\eta} \tag{A.1}$$

$$T_1^2 + T_2^2 = (R_1^2 - R_2^2)^2 \tag{A.2}$$

$$\frac{T_1}{T_2} = \frac{1 - R_1^2}{\lambda\sqrt{1 - \eta}} \tag{A.3}$$

$$T_1^2 T_2^2 = \lambda^2(1 - \eta)(T_1^2 + T_2^2) \tag{A.4}$$

$$R_2^2 T_1^2 + R_1^2 T_2^2 = (R_1^2 - R_2^2)^2 \tag{A.5}$$

$$R_1^2 T_1^2 + R_2^2 T_2^2 = \lambda^2(R_1^2 - R_2^2)^2 \tag{A.6}$$

$$R_1^2 + R_2^2 = \lambda^2 + 1 \tag{A.7}$$

$$(1 - R_1^2)(1 - R_2^2) = -\lambda^2(1 - \eta) \tag{A.8}$$

$$\frac{F_1^2}{R_1} + \frac{F_2^2}{R_2} = \frac{1}{1 - \eta} \tag{A.9}$$

Appendix B

Study of screening properties in the charged Bose gas at $T = 0$

B.1 Dielectric formulation of the many-body problem

B.1.1 Basic definitions

Characterization of the fluid

The system we are going to study is a fluid of charged particles in a neutralizing background. All the results contained in this Appendix concern the zero temperature charged fluid. The fluid is characterized by the adimensional parameter

$$r_s = \frac{r_0}{a_0} \, ,$$

where r_0 is related to the average particle density n through

$$\frac{4\pi}{3} r_0^3 = \frac{1}{n}$$

and therefore has the meaning of the average interparticle distance. The quantity a_0 is the Bohr radius,

$$a_0 = \frac{\hbar^2}{me^2} \, ,$$

m and e being the mass and charge of the charged particle respectively.

The parameter r_s can be identified with the coupling strength. This is immediately seen in the case of a fermion fluid, since r_s is easily seen to be the ratio E_{pot}/E_{kin} of the mean potential-to-mean kinetic energies. For the charged boson fluid at zero temperature a different way is to be followed. The ground state energy $E_{gs}(\lambda)$ as a function of the coupling strength λ is related to the mean potential energy $E_{pot}(\lambda)$ through the relation [PinesNoz66]

$$\frac{dE_{gs}(\lambda)}{d\lambda} = \frac{E_{pot}(\lambda)}{\lambda}.$$

The virial theorem yields

$$\frac{d[r_s^2 E_{gs}(r_s)]}{dr_s} = r_s E_{pot}(r_s).$$

Since E_{gs} and E_{pot} depend only on r_s once they are expressed in Rydberg units $\hbar^2/2ma_0^2$, r_s can be identified with the couplig strength λ. Furthermore the ground state energy is given by

$$E_{gs} = E_0 + \int_0^{e^2} \frac{d\lambda}{\lambda} E_{pot}(\lambda), \tag{B.1}$$

where E_0 is the ground state energy of the non-interacting charged gas. The quantity $E_{xc} = E_{gs} - E_0$ is the exchange and correlation energy. E_{xc} is the mean potential energy plus the kinetic energy excess due to correlations in the particle motion.

Dielectric function approach

The dielectric function $\epsilon(\vec{k}, \omega)$ of a fluid of charged particles constitutes a simple way to study a number of physical properties [MarchTosi84,Tosi94]. $\epsilon(\vec{k}, \omega)$ is defined through the linear response of the fluid to an external electric potential $V_e(\mathbf{r}, t)$ varying in space and time with wave vector $\mathbf{k}$ and frequency ω. Therefore the dielectric function contains information on the dynamical structure of correlations between density fluctuations.

Let $E(\mathbf{r}, t)$ be the field due to the external probe charge $n_e(\mathbf{r}, t)$ plus the induced $\delta n(\vec{r}, t)$ charges and $D(\mathbf{r}, t)$ the correspondent field created in vacuo by the charged probe. The Fourier transformed Poisson equations relate the longitudinal fields to $\epsilon(\vec{k}, \omega)$ through the equation:

$$\frac{1}{\epsilon(\vec{k}, \omega)} = \frac{\mathbf{k} \cdot \vec{E}(\mathbf{k}, \omega)}{\mathbf{k} \cdot \vec{D}(\mathbf{k}, \omega)} = \frac{n_e(\vec{k}, \omega) + n(\vec{k}, \omega)}{n_e(\vec{k}, \omega)} = \frac{V_H(\mathbf{k}, \omega)}{V_e(\mathbf{k}, \omega)}, \tag{B.2}$$

where $n(\vec{k}, \omega)$ is the Fourier transform with respect to space and time of the induced density. Eq.(B.2) defines the Hartree potential V_H. The density induced change $n(\vec{k}, \omega)$ of the system due to the external $V_e(\mathbf{r}, t)$ potential defines the density-density response function $\chi(\vec{k}, \omega)$:

$$n(\vec{k}, \omega) = \chi(\vec{k}, \omega) V_e(\mathbf{k}, \omega). \tag{B.3}$$

Since the response is to be causal and finite, the real and imaginary parts of $\chi(\vec{k}, \omega)$ obey Kramers-Kronig relations. Eqs.(B.2) and (B.3) give the relationship between the dielectric function and the density-density response function:

$$\frac{1}{\epsilon(\vec{k}, \omega)} = 1 + \frac{4\pi e^2}{k^2} \chi(\vec{k}, \omega) \ . \tag{B.4}$$

Finally the moments $M_n(k)$ of order n of the response function are to be defined:

$$M_n(k) = -\int_{-\infty}^{\infty} \frac{d\omega}{\pi} \omega^n Im\chi(\vec{k}, \omega) \ . \tag{B.5}$$

The Fourier transform of the preceding relation tells that the moments are essentially derivatives of $Im\chi(\mathbf{k}, t)$ with respect to the time and hence their determination reduces to the calculation of commutators with the hamiltonian of the system. Furthermore the even order moments vanish as a consequence of the symmetry property $Im\chi(-\omega) = -Im\chi(\omega)$. From Eq.(B.5) is evident that the existence of odd order moments is directly related to the asymptotic behaviour of the response function at high frequencies. It has been shown for fermions [GlickLong71] that only the moments up to the third exist. It can be shown that the same result holds for bosons.

B.1.2   General properties of the dielectric function

Introduction

The dielectric function gives information on single particle and collective excitation spectrum $S(\vec{k}, \omega)$ and, via the fluctuation-dissipation theorem, on the pair correlation function and thermodynamic functions of the fluid. In the static limit it describes the screening of heavy and weakly charged impurities. The analysis of sum rules for density and particle excitations also allow a deep understanding of various dynamic features of the system from a microscopical point view; in fact it has already been noted that the derivation of the different sum rules only requires the knowledge of basic commutation relations with the hamiltonian. In the specific case of a Bose fluid, the symmetry breaking leads to a coupling between density and particle excitations; thus a set of density-density, particle-particle and density-particle sum rules can be derived, allowing a systematic study of the infrared divergences. A summary of the main conclusions on this point is given in the last section of the Appendix. Now we shall focus on the general properties of the dielectric function.

Links between various physical quantities

The response function is intimately related to the dynamic structure factor $S(\vec{k}, \omega)$ of the fluid introduced by van Hove [vanHove54],

$$S(\vec{k}, \omega) = \int\int dr\,dt\,G(r, t)e^{-i(\mathbf{k}\cdot\mathbf{r}-\omega t)} \tag{B.6}$$

$$G(r, t) = \frac{1}{N}\int d\vec{r'}\langle\rho(\vec{r'}, 0)\rho(\vec{r'} + \mathbf{r}, t)\rangle, \tag{B.7}$$

$\rho(\mathbf{r}, t)$ being the particle density operator and N the number of particles.

Since the density-density response function can be written as [MarchTosi84]

$$\chi(\mathbf{r}, t) = -i\theta(t)\langle[\rho(\vec{r'}, 0), \rho(\vec{r'} + \mathbf{r}, t)]\rangle$$

($\theta(t)$ being the step function), it follows that

$$Im\chi(\vec{k}, \omega) = -\frac{N}{2\hbar V}[S(\vec{k}, \omega) - S(-\mathbf{k}, -\omega)]. \tag{B.8}$$

Eq.(B.8) is the fluctuation-dissipation theorem. At zero temperature $S(\vec{k}, \omega) = 0$ for $\omega < 0$ and therefore

$$S(\vec{k}, \omega) = -\frac{2\hbar}{n}Im\chi(\vec{k}, \omega)\theta(\omega) = -\frac{2\hbar k^2}{n4\pi e^2}Im\frac{1}{\epsilon(\vec{k}, \omega)}\theta(\omega). \tag{B.9}$$

The static structure factor measured in diffraction experiments is obtained after integration over energy tranfers:

$$S(k) = \int_{-\infty}^{\infty}\frac{d\omega}{2\pi}S(k, \omega) = -\int_0^{\infty}2\hbar\omega n Im\chi(\vec{k}, \omega). \tag{B.10}$$

Incidentally, Eq.(B.10) is the zeroth-order sum rule for the density-density response function.

The knowledge of the static structure factor allows the determination of the pair correlation function $g(r)$, which describes the spin-averaged probability of finding two particles at a distance r from each other at any given time. In fact $g(r)$ is obtained by Fourier transform of the static structure factor:

$$g(r) = \frac{1}{n}\int\frac{d\mathbf{k}}{(2\pi)^3}e^{i\mathbf{k}\cdot\mathbf{r}}(S(k) - 1).$$

Kimball [Kimball73] and Niklasson [Niklasson74] have independently derived the relation that the pair correlation function must fulfil at short distances, namely

$$g(0) = a_0\left[\frac{dg(r)}{dr}\right]_{r=0}$$

and

$$\lim_{k\to\infty} S(k) = 1 - \frac{6r_s}{r_0^4 k^4} g(0) .$$

The second relation brings a quantitative information on how the short range properties of the fluid structure are connected to the long wave-vector behaviour.

The knowledge of the structure of the fluid allows an evaluation of the mean potential energy E_{pot} per particle:

$$E_{pot} = \frac{n}{2} \int d\mathbf{r} \frac{e^2}{r} [g(r) - 1]$$

and hence of the ground state energy through Eq. (B.1) obtained by means of the virial theorem. From the ground state energy thermodynamic quantities like the pressure P, the isothermal compressibility K_T and the chemical potential μ can be derived by well known relations. For instance:

$$P = n^2 \frac{dE_{gs}}{dn}; \quad \frac{1}{K_T} = n \frac{dP}{dn}; \quad \mu = E_{gs} + \frac{P}{n} .$$

On the other hand the compressibility can be obtained also from the long-wavelength limit of the static dielectric function through the relation:

$$\lim_{k\to 0} \epsilon(k,0) = 1 + v_k n^2 K_T ,$$

where the Fourier transform of the Coulomb interaction $v_k = 4\pi e^2/k^2$ has been introduced. As a consequence of the definition (B.5) together with Kramers-Kronig relations, M_{-1} yields the real part of the static density-density response function:

$$M_{-1}(k) = -\frac{1}{v_k} \left[\frac{1}{\epsilon(k,0)} - 1 \right] .$$

The static dielectric function allows also the determination of the screening charge induced by the presence of a heavy and weakly charged impurity. If the impurity is located at the origin, the screening charge $\rho_s(r)$ is given by

$$\rho_s(r) = \int \frac{d\mathbf{k}}{(2\pi)^3} \left[1 - \frac{1}{\epsilon(k,0)} \right] e^{i\mathbf{k}\cdot\mathbf{r}} .$$

The collective excitation spectrum is given by the frequencies ω_k for which the dielectric function vanishes. Finally the first and third momentum sum rules are to be recalled:

$$M_1(k) = \frac{nk^2}{m} \tag{B.11}$$

$$M_3(k) = \frac{nk^2}{m} \left[\omega_p^2 + \frac{2k^2}{m} E_{kin} + \left(\frac{\hbar k^2}{2m} \right) \right.$$

$$\left. + \frac{\omega_p^2}{N} \sum_{\vec{q}\neq\mathbf{k}} (\hat{\vec{q}} \cdot \hat{\mathbf{k}}) [S(|\mathbf{k} - \vec{q}| - S(q)] \right] , \tag{B.12}$$

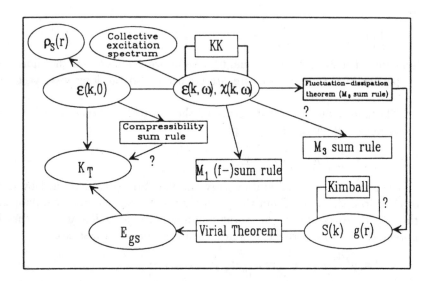

Figure B.1: Summary of the various relationships between the quantities described in the text.

where $\omega_p = (4\pi n e^2/m)^{1/2}$ is the plasma frequency and E_{kin} the mean kinetic energy per particle. Eq.B.11 is also referred to as *f-sum rule*; its derivation involves only the kinetic term of the hamiltonian and therefore it is equal to the first moment of the response function for a non-interacting gas; physically it counts the number of particles irrespective of the interaction among them. The third moment sum rule is composed by different terms. The first is the f-sum rule, the second contains the expectation value of the kinetic energy on the true ground state of the system and therefore brings along some information on the momentum distribution of the particles; finally the third term yields information on the static structure factor.

All the relationships between the various quantities described so far are summurized in Fig.B.1.

B.2   Model dielectric functions

Various approximate model dielectric functions have been set up. To this aim the use of the local field factor concept is crucial. If $\chi_0(k, \omega)$ is the density-density response function of the ideal gas, the induced density change can be written as

$$n(\vec{k}, \omega) = \chi_0(k, \omega)[V_e(k, \omega) + v_k n(\vec{k}, \omega) - v_k G(k, \omega) n(\vec{k}, \omega)] \quad (B.13)$$
$$= \chi_0(k, \omega)\{V_e(k, \omega) + v_k[1 - v_k G(k, \omega)]n(\vec{k}, \omega)\} \quad (B.14)$$

The first term in Eq.(B.13) is due to the external potential and the second to the polarization potential; the sum of the first two is the Hartree potential V_H. The latter term accounts for the density depletion around a given particle, due to the exchange and correlation effects, the so called Coulomb-Pauli hole. The quantity $G(\mathbf{k}, \omega)$ is the local field factor; it prevents the overestimation of correlation effects were only the Hartree potential is considered. The second term in Eq.(B.14) plays the role of an effective potential. The density-density response function $\chi(\vec{k}, \omega)$ then reads:

$$\chi(k, \omega) = \frac{\chi_0(k, \omega)}{1 - v_k[1 - G(k, \omega)]\chi_0(k, \omega)} \ . \tag{B.15}$$

It is worth to point out that Eqs.(B.13) can be justified within the framework of time-dependent density functional theory (DFT) [RungeGross84]. DFT shows that a one-to-one correspondence exists between the applied external potential and the one-body density $n(\vec{r}, t)$. In the theory the exchange-correlation functional $f_{xc}(\mathbf{k}, \omega)$ comes about and this can be identified with $-v_k G(\mathbf{k}, \omega)$.

With respect to a diagrammatic approach, the definition of the proper polarizability $\tilde{\chi}(\mathbf{k}, \omega)$ is useful. It gives the response to the Hartree potential and therefore one has $\chi(\vec{k}, \omega) = \tilde{\chi}(\mathbf{k}, \omega)/\epsilon(\vec{k}, \omega)$. As a first approximation, one could neglect at all the local field factor. In this case the proper polarizability is given by the response function of the ideal gas and the induced density change is driven only by the Hartree potential. This is called Random Phase Approximation and is expected to be sufficiently accurate whenever $r_s < 1$. In the case of the charged Bose fluid at zero temperature, RPA is expected to be inadequate even at weak coupling, since the ideal boson gas is fully condensed in the zero momentum state.

The inclusion of local field effects in a static approximation is the basis of the pioneering work by Singwi, Tosi, Land and Sjölander (STLS) [STLS68]. STLS can be derived from an ansatz for the two-particle Wigner distribution function $f^{(2)}(\vec{r}, \vec{p}; \vec{r'}, \vec{p'}|t)$, in the equation of motion for the one-particle distribution function. The result is a breaking of the hierarchy of equations. The ansatz reads

$$f^{(2)}(\vec{r}, \vec{p}; \vec{r'}, \vec{p'}|t) = f^{(1)}(\vec{r}, \vec{p}|t)f^{(1)}(\vec{r'}, \vec{p'}|t)g(\vec{r} - \vec{r'}) \ .$$

This corresponds to a static approximation for the local field factor:

$$G_{STLS}(k) = -\frac{1}{N} \sum_{\vec{k'}} \frac{\mathbf{k} \cdot \vec{k'}}{k'^2}[S(|\mathbf{k} - \vec{k'}|) - 1] \ . \tag{B.16}$$

Eq.(B.16) has to be self-consistently solved together with Eq.(B.14) and the fluctuation-dissipation theorem. STLS approximation is more suitable than RPA with increasing r_s.

Along the lines of the STLS approximation other static local field models have been proposed. One is due to Vashishta and Singwi (VS) [VS72]; the VS static local field factor is

$$G_{VS}(k) = \left(1 + an\frac{\partial}{\partial n}\right)G_{STLS}(k),$$

where the parameter a has to be self-consistently determined in order to fulfil the compressibility sum rule. Finally the Pathak-Vashishta (PV) [PV73] approximation is built in order to satisfy the third moment sum rule. The local field factor is given by

$$G_{PV} = \frac{1}{(2\pi)^3 n}\int\frac{(\mathbf{k}\cdot\vec{q})^2}{k^2 q^2}[S(|\mathbf{k}-\vec{q}|) - S(q)]d\vec{q}.$$

The low k-vector behaviour ($kr_s a_0 \ll 1$) of the static local field factor can be shown to be

$$G(k,0) = \gamma k^2 = \frac{1}{4\pi n^2 e^2}\left[\frac{1}{K_0} - \frac{1}{K}\right]k^2, \tag{B.17}$$

where the second equality defines γ and K_0 and K are the compressibilities of the free and interacting gas, respectively. Eq.(B.13) is a consequence of the definition of the local field factor (see e.g. Eq.(B.15)) and of the compressibility sum rule. Since exchange and correlation add to the Coulomb repulsion an effective attractive interaction between the particles and therefore γ is expected to be positive.

B.3   Application to charged bosons

B.3.1   Introduction

The dielectric function approach has been widely used to study the fluid of charged fermions as a model for conduction electrons in normal metals and for electron-hole liquid in semiconductors, to fluids in the classical regime as models for ionic liquids and molten salts [SingTosi81]. At variance with the fermion case, little attention has been devoted to the charged boson fluid. The charged Bose gas has been studied in a pioneering papaer by Foldy [Foldy61] within the Bogoliubov formalism. Hore and Frankel [HoreFrank75] have given a full evaluation of $\epsilon(\vec{k},\omega)$ at any temperature within the Random Phase Approximation. STLS for charged bosons has been studied by Caparica and Hipolito [CapHip82]. Recently Quantum Monte Carlo simulations for the static response function [Alder92] and for the momentum distribution function [Moroni96] have been available. Earlier data are due to Hansen and Mazighi [Hansen78] and to Ceperley and Alder [CepAlder80]. These latter reveals that the charged Bose gas is weakly coupled only at $r_s << 1$ and undergoes a Wigner crystallization at $r_s \simeq 160$. This value has to be contrasted with the correspondent value for fermions, which is 80 about. The large difference still constitutes a puzzling feature, if one realizes the fact that at large r_s values the effects of the statistics are expected to be unimportant. The analysis of charged boson fluids would help to disentangle the effects due to the statistics of the particles from the effects due to Coulomb correlations. Finally charged bosons fluids can be also relevant in systems of astrophysical interest like white dwarf stars. Beyond the relationship to superconductivity, all the previous considerations too motivate the present study of charged boson fluids.

B.3.2   Analytical relations

The application of the model dielectric function approach to charged bosons is in principle easier with respect to the fermion case, because some work can be carried out analyticaly. This possibility basically stems from the simple form that the free density-density response function $\chi_0(\mathbf{k}, \omega)$ assumes in the case of bosons [HoreFrank75] with respect to the hard-to-use Lindhard function for fermions [Mahan]:

$$\chi_0(k, \omega) = \frac{nk^2}{m} \frac{1}{\omega(\omega + i\eta) - (\hbar k^2/2m)} , \qquad (B.18)$$

η being infinitesimal and positive. The excitation spectrum is given by

$$\omega_k = \sqrt{\omega_p^2[1 - G(k)] + \left(\frac{\hbar k^2}{2m}\right)^2} . \qquad (B.19)$$

The preceding equation evidences that the small k behaviour of the plasma frequency is driven by the correspondent behaviour of the static structure factor, whereas at high k vectors the excitation spectrum correctly assumes the free particle energy.

The static structure factor is given by

$$S(k) = \frac{\hbar k^2}{2m\omega_k} = \frac{1}{\sqrt{1 + \frac{12r_s}{(kr_0)^4}[1 - G(k)]}} \quad ; \qquad (B.20)$$

the first equality constitutes the Feynman relation.

Finally it is important to recall that the compressibility of the ideal Bose gas is zero due to fully condensation of the particles in zero momentum state. This fact has striking consequences on the properties of the charged boson gas. In fact it has been already noted that the quantity γ in Eq.(B.17) is positive and therefore the compressibility K of the interacting Bose gas turns out to be negative at all r_s.

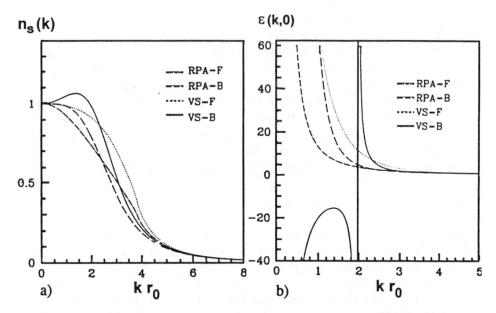

Figure B.2: (a) Fourier transform $n_s(k)$ of the screening charge density around a static, pointlike unit charge as a function of kr_0 and at $r_s = 5$. Different curves refer to RPA and to VS for bosons (B) and fermions (F). (b) The correspondent static dielectric function.

B.3.3   Summary of the main conclusions

Here the main conclusion on the screening properties of the charged Bose gas in different static local field theories are summarized.

Static screening

Fig.B.2a displays the Fourier transform $n_s(k)$ of the screening charge density around a static, pointlike unit charge as a function of kr_0 and at $r_s = 5$. Different curves refer to RPA and to VS for bosons and fermions. The correspondent static dielectric function is given in Fig.B.2b. Differences in static screening between bosons and fermions arise in a low-to-moderate coupling regime. The static screening charge for fermions shows a weak singularity at $k = 2k_F \simeq 3.8/r_0$, k_F being the Fermi wavevector; this is a fingerprint of the discontinuity in the momentum distribution at the Fermi level and manifests in existence of Friedel oscillations in real space. On the other hand, at the same r_s value the boson fluid is characterized by a screening charge larger than the external one in a restricted range of kr_0 values. This range corresponds to a negative static dielectric function. Since the ground state energy of the boson fluid is enterily due to correlations, a local pile-up of screening charge around

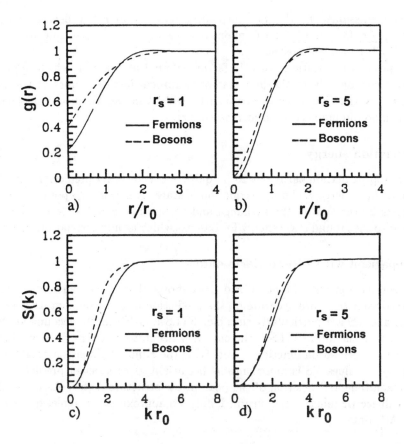

Figure B.3: Pair correlation function $g(r)$ and static structure factor $S(k)$ for $r_s = 1, 5$ within the STLS approximation for charged Bose fluid.

the impurity may take place. This is usually referred to as overscreening effect and comes about as far as local corrections are taken into account. Indeed the RPA boson screening charge does not show such a feature. Even though it is not explicitly shown in the figure, this effect is important even for small r_s values, since it is intimately related to the negative compressibility of the charged Bose fluid.

Structure

Fig.B.3 shows the pair correlation function and the static structure function for $r_s = 1, 5$ as calculated within the STLS model. Differences in the structure between fermions and bosons are evident at low coupling and tend to disappear already at $r_s = 5$. The differences are mainly restricted to short distances.

The value assumed by the pair correlation function at $r = 0$ decreases with increasing r_s from the $1/2$ for fermions because of the Pauli principle and from the $g(0) = 1$ for ideal bosons.

A peak in the static local field factor is found at $kr_0 \simeq 4$, which approximately corresponds to the first star of reciprocal lattice vectors of the bcc Wigner crystal. The peak is particularly evident in the VS local field factor and is more pronounced with increasesing r_s.

Correlation energy

STLS approximation turns out to be quite accurate in its predictions on the ground state energy and on the equation of state. It is worth to point out that for $r_s \simeq 20$ the sum of the exchange and correlation energies for fermions is similar to the ground state energy for bosons (which is only correlation energy).

Comparison with Monte Carlo data

The comparison with Monte Carlo data shows that all of the cited model dielectric functions are accurate on the prediction of the ground state energy. STLS and VS are particularly suitable. STLS and VS nicely account for the structure of the fluid. STLS is the only very accurate model in the low r limit of the pair correlation function, it satisfying the Kimball-Niklasson's relations. At large r_s values VS is more suitable because it gives a better prediction on the magnitude and position of the peak in the structure factor. As far as the static dielectric function is concerned, only VS approximation agrees quite well with MC data.

Excitation spectrum

Fig.B.4 displays the plasma frequency $\omega_p(k)$ normalized to ω_p as a function of kr_0 for fermions and bosons at different r_s values within STLS approximation. A negative dispersion coefficient at all r_s characterizes the boson excitation spectrum, whereas a critical r_s value exists for fermions. The negative dispersion coefficient is a characteristic of strong correlations. In fact as far as r_s and hence the correlations increase, collective excitations in the fluid are most energetically favourite at a finite wave-vector. The difference between fermions and bosons is due once again to the absence of a kinetic energy term in case of bosons and is connected with the negative value of the static dielectric function (and of the compressibility) in a restricted range of kr_0 values.

B.3.4   Sum rule approach

Density-density sum rules have been extensively used in charged fermion fluids to show that the plasmon is the only relevant long wavelength excitation,

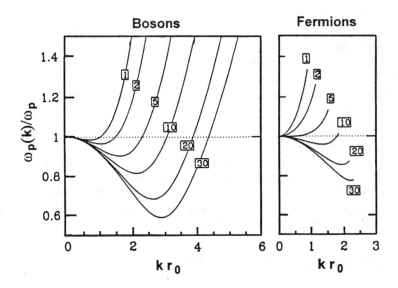

Figure B.4: Excitation spectrum $\omega_p(k)$ normalized to ω_p as a function of kr_0 for bosons (left) and fermions (right) at different r_s within the STLS approximation.

since it exhausts the f-sum rule. In neutral systems a similar analysis has been carried out by Feynman [Feynman54] for the long wavelength phonon excitation spectrum. A complete discussion of density-density, particle-particle and density-particle sum rules in ^{4}He has been given by Stringari [Giorgini90,Giorgini92,Stringari92]. A systematic study of the charged Bose fluid within a sum rule approach is given in Ref. [Chiofalo96]. Here we wish to summarize the main comments and results.

First, the interpretation of the sum rules in the charged Bose gas requires that in the long wavelength limit the single particle excitation spectrum coincides with the density excitation spectrum, namely the plasmon collective excitation. Foldy [Foldy61] has proved the previous statement in the high density limit. A general proof at any density hinges on the Feynman's argument [Feynman54], originally referred to long wavelength phonons in ^{4}He. The argument notices that the lowest lying excited states are those which involve large groups of atoms and therefore long wavelengths; the rearrangements of individual atoms among themselves do not change the boson wavefunction because of the statistics. Feynman's argument only uses the symmetry of the boson wavefunction and therefore applies irrespective of the existence of long range interactions. Thus to leading order the lowest excited states in a Bose fluid are density waves: phonon excitations in the neutral system and plasmons in the charged one. A further argument is due to Gavoret and Nozières [GavoNoz64], who note that in the presence of a condensate the single-particle and the collective excitaions have the same spectrum apart from a factor. The existence of a gap in the single particle excitation spectrum together with the hypothesis of non-interacting excitations allows a description of the long-wavelength physics in terms of the Quantum Hydrodynamics (QH), in which the elementary excitations are described by an effective free boson hamiltonian. The description in terms of QH is complementary to the sum rule approach.

Second, it can be shown that the plasmon exhausts the density-density sym rules to leading order in k. The proof is based only on the analysis of the M_{-1}, M_1 and M_3 moments of the density-density response function. In contrast, in the corresponding neutral systems the phononic and multiparticle contributions to the third moment are both of order k^4 [Stringari92].

Some upper bounds on plasmon dispersion can be derived from the density-density sum rules [Stringari92,Chiofalo95]

$$\omega_k^{(min)} \leq \frac{M_0(k)}{M_{-1}(k)} \leq \sqrt{\frac{M_1(k)}{M_{-1}(k)}} \leq \frac{M_1(k)}{M_0(k)} \leq ... \tag{B.21}$$

All of the above upper bounds imply the relation $d^2\omega_k/dk^2|_{k=0} < 0$ for the collective excitation spectrum. This is intimately connected with the negative compressibility of the charged Bose gas.

Consider the Hugenholtz and Pines relation [HugenPines59]

$$\mu = \int \frac{d\vec{q}}{(2\pi)^3} v_q F_1(q) \quad ;$$

this links the chemical potential to the off-diagonal two-body density matrix which enters through the function $F_1(q)$ (for details and definitions, see the included reprinted paper [Chiofalo96]). It can be shown that the Hugenoltz and Pines relation is equivalent to state that the long wavelength limit of the single particle excitation energy is the plasmon energy.

Finally, the long range nature of Coulomb interactions in the charged Bose fluid implies sharper infrared divergences than in the analogous neutral system. A summary of the long wavelength behaviour of the density-density, density-particle and particle-particle excitations is given in Tab.B.1.

Table B.1: Leading and subleading $k \to 0$ contributions to various matrix elements and sum rules. The single–plasmon and multiparticle contributions are distinguished. ρ_k is the density operator and a_k, $a_k^\dagger$ the particle operators "$\leq k^2$" means "of order k^2 or lower", "(*)" indicates that strong cancellations occur and • marks the f-, the Bogoliubov and the commutation relation sum rules which are identical in the charged and in the neutral Bose fluids.

	Plasmon	Multiparticle	
$\langle n|\rho_k|0\rangle = \langle n|\rho_k^\dagger|0\rangle$	$\sqrt{\dfrac{Nk^2}{2m\omega_p}} \propto k$	$\propto k^2$	
$\langle n|a_k|0\rangle$	$-\sqrt{n_0\dfrac{\omega_p}{4\varepsilon_k}} \propto \dfrac{1}{k}$	$\sqrt{\dfrac{n_0}{N}\dfrac{m\omega_p}{2|q||k-q|}} \propto \dfrac{1}{\sqrt{k}}$	
$\langle n|a_k^\dagger|0\rangle$	$\sqrt{n_0\dfrac{\omega_p}{4\varepsilon_k}} \propto \dfrac{1}{k}$	$\sqrt{\dfrac{n_0}{N}\dfrac{m\omega_p}{2|q||k-q|}} \propto \dfrac{1}{\sqrt{k}}$	
ω_{n0}	ω_p	$2\omega_p$	
$\sum_n |\langle n|\rho_k|0\rangle|^2\omega_n^{-1}$	$N\varepsilon_k\omega_p^{-2}$	$\propto k^4$	
$\sum_n |\langle n|\rho_k|0\rangle|^2$	$N\varepsilon_k\omega_p^{-1}$	$\propto k^4$	
$\sum_n |\langle n|\rho_k|0\rangle|^2\omega_n$	$N\varepsilon_k$	$\propto k^4$	•
$\sum_n |\langle n|\rho_k|0\rangle|^2\omega_n^2$	$N\varepsilon_k\omega_p$	$\propto k^4$	
$\sum_n |\langle n|\rho_k|0\rangle|^2\omega_n^3$	$N\varepsilon_k\omega_p^2$	$\propto k^4$	
$\sum_n \left[|\langle n|a_k|0\rangle|^2 + |\langle n|a_k^\dagger|0\rangle|^2\right]\omega_n^{-1}$	n_0m/k^2	$m^2n_0\omega_p/64\rho k$	
$\sum_n |\langle n|a_k|0\rangle|^2$	$n_0m\omega_p/2k^2$	$m^2n_0\omega_p^2/64\rho k$	
$\sum_n \left[|\langle n|a_k|0\rangle|^2 - |\langle n|a_k^\dagger|0\rangle|^2\right]$	$\leq$const (*)	$\leq$const (*)	•
$\sum_n \left[|\langle n|a_k|0\rangle|^2 + |\langle n|a_k^\dagger|0\rangle|^2\right]\omega_n$	$n_0m\omega_p^2/k^2$	$m^2n_0\omega_p^3/16\rho k$	
$\sum_n |\langle n|a_k|0\rangle|^2\omega_n$	$n_0m\omega_p^2/2k^2$	$m^2n_0\omega_p^3/32\rho k$	
$\sum_n \langle n|\rho_k|0\rangle\langle n|a_k + a_k^\dagger|0\rangle/\omega_n$	$\leq k^2(*)$	$\leq k^2(*)$	
$\sum_n \langle n|\rho_k|0\rangle\langle n|a_k - a_k^\dagger|0\rangle$	$-\sqrt{Nn_0}$	$\leq k^2(*)$	•
$\sum_n \langle n|\rho_k|0\rangle\langle n|a_k|0\rangle$	$-\sqrt{Nn_0}/2$	$\propto k^{3/2}$	
$\sum_n \langle n|\rho_k|0\rangle\langle n|a_k + a_k^\dagger|0\rangle\omega_n$	$\leq k^2(*)$	$\propto k^{3/2}$	
$\sum_n \langle n|\rho_k|0\rangle\langle n|a_k|0\rangle\omega_n$	$\sqrt{Nn_0}\omega_p/2$	$\propto k^{3/2}$	

Bibliography

[Adamowski89] J. Adamowski, Phys. Rev. **B39**, 3649 (1989); Acta Physica Polonica **A75**, 51 (1989).

[Alder92] G. Sugiyama, C. Bowen and B. J. Alder, Phys. Rev. **B46**, 13042 (1992).

[AlexKrebs92] A. S. Alexandrov and A. B. Krebs, *Polarons in high-temperature superconductors*, Sov. Phys. Usp. **35**, 5 (1992).

[AlexMott93] A.S. Alexandrov and N.F. Mott, Superc. Sci. Technol. **6**, 215 (1993).

[AlexMott94a] A. S. Alexandrov and N. F. Mott, *Bipolarons*, Reports on Progress in Physics **57** (1994).

[AlexMott94b] A. S. Alexandrov and N. F. Mott, *High Temperature Superconductors and other Superfluids*, Taylor and Francis, London (1994).

[AlexRann81a] A. S. Alexandrov and J. Ranninger, Phys. Rev. **B23**, 1796 (1981).

[AlexRann86] A. S. Alexandrov, J. Ranninger and S. Robaszkiewicz, Phys. Rev. **B33**, 4526 (1986).

[AlexRann81b] A. S. Alexandrov and J. Ranninger, Phys. Rev. **B24**, 1164 (1981).

[Anderson75] P. W. Anderson, Phys. Rev. Lett. **34**, 953 (1975).

[Anderson87a] P.W. Anderson, Science **235**, 1196 (1987).

[Anderson87b] P.W. Anderson, G.Baskaran, Z. Zou and T. Hsu, Phys. Rev. Lett. **58**, 2790 (1987).

[Arai92] M. Arai *et al.*, Phys. Rev. Lett. **69**, 359 (1992).

[AshMerm] N. W. Ashcroft and N. Mermin, *Solid State Physics*, HRW International Editions, Hong Kong (1976).

[Athan95] N. Athanassopoulou and J. R. Cooper, private comm. (1995).

[Barton71] G. Barton, Reports on Progr. in Phys., **3**, 1888 (1971)

[Bassani91] F. Bassani, M. Geddo, G. Iadonisi and D. Ninno, Phys. Rev. **B43**, 5296, (1991).

[Bazhenov89] A. V. Bazhenov and V. B. Timofeev, Physica **C162-164**, 1247 (1989).

[BCS57] J. Bardeen, L. Cooper and J. R. Schrieffer, Phys. Rev. **108**, 1175 (1957).

[BednMuller88] J. G. Bednorz and K. A. Z. Müller, Rev. Mod. Phys. **60**, 585 (1988).

[BednMuller86] J. G. Bednorz and K. A. Z. Müller, Phys. Rev. **B64**, 189 (1986).

[Berghuis90] P. Berghuis, P. H. Kes, G. M. Stollman and J. van Bentum, Physica **C167**, 348 (1990).

[Blatt64] J. M. Blatt, *Theory of superconductivity*, Academic Press, New York, London (1964).

[BuchWach92a] B. Bucher and P. Wachter, Helv. Phys. Acta **65**, 387 (1992).

[BuchWach92b] B. Bucher, J. Kapinski, E. Kaldis and P. Wachter, Phys. Rev. **B45**, 3026 (1992).

[BuchWach95] B. Bucher and P. Wachter, Phys. Rev. **B51**, 3309 (1995).

[Burns92b] G. Burns, *High-Temperature Superconductivity. An introduction*, Academic Press, Boston (1992).

[Calvani96] For a review see: P. Calvani, "Infrared properties of High-T_c Cuprates", in *High-T_c Superconductivity: Theory and Experiment*, ed. by M. Acquarone, World Scientific, Singapore (1996) and references therein.

[CapHip82] A. A. Caparica and O. Hipolito, Phys. Rev. **A26**, 2832 (1982).

[Capone94] G. Capone, V. Cataudella, M. L. Chiofalo, R. Di Girolamo, G. Iadonisi, F. Liguori and D. Ninno, "Theory of dynamical screening effects in exciton and bipolaron formation; an application to strongly photoexcited semiconductors and to the bipolaron model for high-T_c superconductivity", Proc. of the International Workshop on superconductivity and strongly correlated electron systems, held in Amalfi, 14-16 October 1993, World Scientific Ed., Singapore (1994).

[Cardona94] A. P. Litvinchuk, C. Thomsen and M. Cardona, "Infrared-active vibrations of high-temperature superconductors: experiment and theory", in *Physical Properties of High Temperature Superconductors IV*, Ed. by D. M. Ginsberg, World Scientific Singapore (1994) and references therein.

[Carrington93] A. Carrington, D. J. Walker, A. P. Mackenzie and J. R. Cooper, Phys. Rev. **B48**, 13051 (1993).

[Cataud91] V. Cataudella, G. Iadonisi and D. Ninno, Physica Scripta **T39**, 71 (1991).

[Cataud92] V. Cataudella, G. Iadonisi and D. Ninno, Europhys. Lett. **17**, 709 (1992).

[Cataud96] V. Cataudella, G. Iadonisi, D. Ninno and M. L. Chiofalo, "On the boson-fermion model for high-T_c superconductivity", submitted.

[CepAlder80] D. M. Ceperley and B. J. Alder, J. Physique Coll. **C7**, 295 (1980).

[Chak87] B.K. Chakravarty, D. Feinberg, Z. Hang and M. Avignon, Solid State Commun. **64**, 1147 (1987).

[Chevrel-note] Chevrel-phases are compounds like $REMo_6X_8$ with RE= Rare Earth element and X=S, Se and are characterized by the fact that while the temperature is lowered they enter a normal-superconductor transition ($T_c \simeq 2K$) and then a superconductor to normal ferromagnetic metal state transition at a lower temperature ($T_c < 1K$). $Ba(Pb_{1-x}Bi_x)O_3$ and $SrTiO_{3-x}$ as well are other oxides (but not copper-based) with critical temperatures up to 13K. Heavy-electron (usually f-electron) metals are UPt_3, UBe_{13} and are characterized by large electron effective masses, as determined from normal-state electronic specific heat measurements; other heavy-electron metals like $CeAl_3$, $CeCu_6$ belong to the same class but are not superconducting. BEDT are organic materials based on bis(ethylenedithia)tetrathiafulvalene or ET and have T_c from 1 up to about 11K and are almost unidimensional in the transport properties. All these materials have been discovered before La_2CuO_4 was.

[Chiofalo92] G. Iadonisi, M. L. Chiofalo, V. Cataudella and D. Ninno, "Mobility of Biplasmapolarons and High-T_c Superconductivity", Il Nuovo Cimento **D15**, 1035 (1993) and references therein **17**, 709 (1992).

[Chiofalo94] M. L. Chiofalo, S. Conti and M.P.Tosi, "Dielectric screening in charged Bose versus Fermi liquids", Mod. Phys. Lett. **B8**, 1207-1221 (1994).

[Chiofalo95] M. L. Chiofalo, S. Conti, S. Stringari and M.P.Tosi, "Upper bounds on plasmon dispersion in the degenerate boson plasma", J. Phys.: Condensed Matter 7, L85 (1995).

[Chiofalo96] M. L. Chiofalo, S. Conti and M. P. Tosi, "Sum rules for density and particle excitations in a superfluid of charged bosons" J. Phys.: Condensed Matter, in press.

[Cohen] C. Cohen-Tannoudji, B. Diu and F. Laloë, Quantum Mechanics, Vol. I, Wiley, New York (1977).

[Conti94] S. Conti, M. L. Chiofalo and M.P.Tosi, "Dielectric response of the degenerate plasma of charged bosons in static local-field approximations" J. Phys.: Condensed Matter 6, 8975 (1994).

[Cooper88] S. L. Cooper, M. V. Klein, B. G. Pazol, J. P. Rice and D. M. Ginsberg, Phys. Rev B37, 5920 (1988).

[Crawford90] M. K. Crawford et al., Science 250, 1390 (1990).

[Degiorgi89] L. Degiorgi, S. Rusiecki and P. Wachter, Physica C161, 239 (1989).

[DeJongh88] L.J. De Jongh, A comparative study of (bi)polaronic (super)conductivity in high-and low-T_c superconducting oxides, Physica C152, 171 (1988).

[Dessau93] D. S. Dessau et al., Phys. Rev. Lett. 71, 2781 (1993).

[Devreese77] L. F. Lemmens, J. T. Devreese and F. Brosens, Physica Status Solidi B82, 439 (1977).

[Devreese84] J. T. Devreese, in Polarons and Excitons in Polar Semiconductors and ionic Crystals, J. T. Devreese and F. Peeters Eds., Plenum New York (1984).

[Devreese91] G. Verbist, F.M. Peeters and J. Devreese, Phys. Rev. B43, 2712 (1991).

[Devreese95] J. T. Devreese, G. Verbist and F. M. Peeters, in Proceedings of the Workshop on Polarons and Bipolarons in High-T_c Superconductors and related Materials, Cambridge University Press (1995).

[Devreese96] J. T. Devreese, "Polarons" in Encyclopedia of Applied Physics, Vol.14. VCH Publishers (1996).

[Devreese96] J. T. Devreese and Smondyrev, Sol. St. Comm. 98, 1091 (1996).

[Eliash60] G. M. Eliashberg, Sov. Phys. JETP **11**, 696 (1960).

[Emin89] D. Emin, Phys. Rev. Lett. **62**, 1544 (1989).

[Emin93] D. Emin, Phys. Rev. **B48**, 13691 (1993).

[Evrard72] R. Evrard, in "Polarons in Ionic Crystals and Polar Semiconductors", J.T. Devreese Ed., North-Holland, Amsterdam (1972), pag. 29.

[EminHill89] D. Emin and M.S. Hillery, Phys. Rev. **B39**, 6575 (1989).

[Dynes94] R. C. Dynes, Sol. State Comm. **92**, 53 (1994).

[Fehr85] G. W. Fehrenbach, W. Schäfer and R. G. Ulbrich, J. of Luminescence **30**, 154 (1985).

[FetterWalecka] A. L. Fetter and J. D. Walecka, "Quantum Theory of Many-Particle systems" (McGraw-Hill, 1971), Chap. 4.

[Feynman54] R. P. Feynman, Phys. Rev. **94**, 262 (1954).

[Feynman62] R. P. Feynman, R. W. Hellwarth, C. K. Iddings and P. M. Platzmann, Phys. Rev. **127**, 1004 (1962).

[Foldy61] L. L. Foldy, Phys. Rev. **124**, 649 (1961).

[Franck91] J. P. Franck *et al.*, Physica **C185-189**, 1379 (1991).

[Franck94] For a review: J. P. Franck, "Experimental studies of the isotope effect in high temperature superconductors" in *Physical Properties of High Temperature Superconductors IV*, Ed. by D. M. Ginsberg, World Scientific Singapore (1994).

[Friedl90] B. Friedl, C. Thomsen and M. Cardona, Phys. Rev. Lett. **65**, 915 (1990).

[FriedLee89a] R. Friedberg and T. D. Lee Phys. Lett. **A138**, 423 (1989).

[FriedLee89b] R. Friedberg and T. D.Lee Phys. Rev. **B40**, 6745 (1989).

[FriedLee90] R. Friedberg, T.D. Lee and H.C. Ren Phys. Rev. **B42**, 4122 (1990).

[Friedman90] T.A. Friedmann, M.W. Rabin, J. Giapintzakis, J.P. Rice and D.M. Ginsberg, Phys. Rev. **B42**, 6217 (1990).

[Froh50] H. Fröhlich, H. Pelzer and S. Zienau, Phylos. Mag. **41**, 221 (1950).

[Froh63] H. Fröhlich, in *Polarons and Excitons*, Eds. C. G. Kuper and A. Whitfield (Oliver and Boyd, Edinburgh, 1963), pag. 1.

[Gajic92] R. Gajic, E. H. Salje, Z. V. Popovic and H. L. Dewing, J. Phys.: Condensed Matter **4**, 9643 (1992).

[GavoNoz64] J. Gavoret and P. Nozières, Ann. Phys. **28**, 349 (1964).

[Gay72] J. G. Gay, Phys. Rev. **B4**, 2567 (1971); Phys. Rev. **B6**, 4884 (1972).

[Gersten88] J. I. Gersten, Phys. Rev. **B37**, 1616, (1988).

[Giaever60] I. Giaever, Phys. Rev. Lett. **5**, 147 464 (1960).

[Giannozzi96] P. Giannozzi and W. Andreoni, Phys. Rev. Lett., in press.

[GinzLandau50] V. L. Ginzburg and L. D. Landau, Zh. Eksp. Teor. Fiz. **20**, 1064 (1950).

[Giorgini90] S. Giorgini and S. Stringari, Physica **B165-166**, 511 (1990).

[Giorgini92] S. Giorgini, L. Pitaevskij and S. Stringari, Phys. Rev. **B46**, 6374 (1992).

[GlickLong71] A. J. Glick and W. F. Long, Phys. Rev. **B4**, 3455 (1971).

[Gough87] C. E. Gough *et al.*, Nature **326**, 855 (1987).

[Hackl89] R. Hackl, R. Kaiser, W. Gläser, Physica **C162-164**, 431 (1989).

[Hansen78] J. P. Hansen and R. Mazighi, Phys. Rev. **A18**, 1282 (1978).

[HaugSchmitt84] H. Haug and S. Schmitt-Rink, *Electron Theory of the Optical Properties of Laser-Excited Semiconductors*, Progr. Quant. Electon. **9**, 3 (1984).

[Hopf58] J. J. Hopfield, Phys. Rev. **112**, 1555 (1958)

[HoreFrank75] S. R. Hore and N. E. Frankel, Phys. Rev. **B12**, 2619 (1975).

[HugenPines59] N. Hugenholtz and D. Pines, Phys. Rev. **116**, 489 (1959).

[Iadonisi84] G. Iadonisi, *Electron-Phonon Interaction: Effects in the Excitation Spectrum of Solids*, La Rivista del Nuovo Cimento **7** n. 11 (1984).

[Iadonisi85] G. Iadonisi and V. Marigliano Ramaglia, Il Nuovo Cimento **D6**, 193 (1985).

[Iadonisi89] G. Iadonisi, F. Bassani and G. C. Strinati, Physica Status Solidi **153**, 611 (1989).

[Iadonisi93] G. Iadonisi, M. L. Chiofalo, V. Cataudella and D. Ninno, "Plasmon-phonon cooperative effects in the dilute large bipolaron gas: a possible mechanism for high-T_c superconductivity", Phys. Rev. **B48**, 12966 (1993).

[Iadonisi95] G. Iadonisi, V. Cataudella, D. Ninno and M. L. Chiofalo, "Polaron and Bipolaron Coexistence in High-T_c Superconductivity", Phys. Lett. **A196**, 359 (1995).

[Iadonisi96a] G. Iadonisi, G. Capone, V. Cataudella and D. Ninno, to appear in Physica Status Solidi (1996).

[Iadonisi96b] G. Iadonisi, G. Capone, V. Cataudella and G. De Filippis, "Electron screening effects on self-trapping of polarons", to appear in Phys. Rev. **B** (1996).

[Ito91a] T. Ito, H. Takagi, S. Ishibashi, T. Ido and S. Uchida, Nature **350**, 596 (1991).

[Ito91b] T. Ito, Y. Nakamura, H. Takagi and S. Uchida, Physica **C185-189**, 1267 (1991).

[Iye92] For a review, see: Y. Iye, "Transport properties of high T_c cuprates", in *Physical Properties of High Temperature Superconductors III*, Ed. by D. M. Ginsberg, World Scientific Singapore (1992) and references therein.

[Josephson62] B. D. Josephson, Phys. Rev. Lett. **1**, 251 (1962).

[Kamaras90] K. Kamaras, S. L. Herr, C. D. Porter, N. Tache, D. B. Tanner, S. Etemad, T. Venkatesen, E. Chase, A. Inam, X. D. Wu, M. S. Hegde and B. Dutta, Phys. Rev. Lett. **64**, 84 (1990).

[Kane78] E. O. Kane, Phys. Rev. **B18**, 6849 (1978).

[Kimball73] J. C. Kimball, Phys. Rev. **A7**, 1648 (1973).

[Kirtley90] J. R. Kirtley, Int. J. of Mod. Phys. **B4**, 201 (1990).

[Kresin88] V. K. Kresin and H. Morawitz, Phys. Rev. **B37**, 7854, (1988).

[Krusin89] L. Krusin-Elbaum *et al.*, Phys. Rev. Lett. **62**, 217 (1989).

[Kubo91] Y. Kubo, Y. Shimakawa, T. Manako and H. Igarashi, Phys. Rev. **B43**, 7875 (1991).

[Kuper63] For a review see: "Polarons and Excitons", edited by G.C. Kuper and G.D. Whitfield (Plenum, New York, 1963) and references therein.

[Landau53] L. D. Landau, Z. Phys. **3**, 664 (1953).

[LandauPekar46] L. D. Landau and S. I. Pekar, Zh. Eksp. Teor. Fiz. **16**, 341 (1946).

[LLP53] T. D. Lee, F. E. Low and D. Pines, Phys. Rev. **90**, 297 (1953).

[Loram93] J. W. Loram, K. A. Mirza, J. R. Cooper and W. Y. Liang, Phys. Rev. Lett. **71**, 1740 (1993).

[Lupi92] S. Lupi, P. Calvani, M. Capizzi, P. Maselli, W. Sadowski and E. Walker, Phys. Rev. **B45**, 12470 (1992).

[Machi91] T. Machi, I. Tomeno, T. Miyatake and S. Tanaka, Physica **C173**, 32 (1991).

[Mackenzie93] A. P. Mackenzie, S. R. Julian, G. G. Lonzarich, A. Carrington, S. D. Hughes, R. S. Liu and D. C. Sinclair, Phys Rev. Lett. **71**, 1238 (1993).

[Mahan] G. D. Mahan, *Many-Particles Physics*, Plenum Press 1981, chap. 6.

[MarchTosi84] N. H. March and M. P. Tosi, *Coulomb Liquids*, Academic Press London (1984).

[Marel92] D. van der Marel and G. Rietveld, Phys. Rev. Lett. **69**, 2575 (1992).

[MattBard58] D. C. Mattis and J. Bardeen, Phys. Rev. **111**, 412 (1958).

[MeissOchs33] Meissner and Ochsenfeld, Naturwiss **21**, 787 (1933).

[Micnas90] R. Micnas, J. Ranninger and S. Robaszkiewicz, Rev. of Mod. Phys. **62**, 113 (1990).

[Micnas95] R. Micnas, M. H. Pedersen, S. Schafroth and T. Schneider, Phys. Rev. **B52**, 16223 (1995).

[Migdal58] A. B. Migdal, Sov. Phys. JETP **7**, 996 (1958).

[Mihail90] D. Mihailovic, C. M. Foster, K. Voss and J. Heeger, Phys. Rev. **B42**, 7989 (1990).

[Miyakawa89] N. Miyakawa, D. Shimada, T. Kido and N. Touda, J. Phys. Soc, Jpn. **58**, 383 (1989).

[Mooradian66] A. Mooradian and G. B. Wright, Phys. Rev. Lett. **16**, 999 (1966).

[Mooradian72] For a review: A. Mooradian, "Raman Spectroscopy of Solids", in *Laser Handbook* Vol.2 Part E, Ed. by F. T. Arecchi and E. O. Schulz-Dubois, North-Holland Pub. Co. Amsterdam (1972).

[Moroni96] S. Moroni, S. Conti and M. P. Tosi, Phys. Rev. B, in press.

[Mott93] N.F. Mott, J. Phys.: Condens. Matter 5, 3487 (1993) and references therein.

[Nasu87] K. Nasu, Phys. Rev. B35, 1748 (1987).

[Nieder93] Ch. Niedermayer, C. Banhord, U. Binninger, and H. Glückler Phys. Rev. Lett. 71, 1764 (1993).

[Niklasson74] G. Niklasson, Phys. Rev. B10, 3052 (1974).

[Ninno94] D. Ninno, V. Cataudella, G. Iadonisi and F. Liguori, J. Phys.: Cond. Matter 6, 9335 (1994).

[Noh88] T.W. Noh et al, Phys. Rev. B36, 8866 (1988).

[Note] The concept of "plasmaron" was first introduced by Overhauser and indicate the quasi-particle which results from the electron-plasmon interaction we are considering.

[NozSchmitt85] P. Nozieres and S. Schmitt-Rink, Jour. Low Temp. Phys. 59, 195 (1985).

[Ogg46] R. A. Ogg, Phys. Rev. 69, 243 (1946).

[Olson90] C. G. Olson *et al.*, Solid State Commun. 76, 411 (1990).

[Osofsky93] M. S. Osofsky *et al.*, Phys. Rev. Lett. 71, 2315 (1993).

[Overend94] N. Overend, M. A. Howson and I. D. Lawrie, Phys. Rev. Lett. 72, 3238 (1994).

[Overhauser71] A. W. Overhauser, Phys. Rev. B3, 1888 (1971)

[Pickett89] For a review: W. Pickett, Rev. Modern Phys. 61, 433 (1989).

[PietroStrass92] L. Pietronero and S. Strässer, Europhys. Lett. 18, 627 (1992).

[Pines63] D. Pines, *Elementary Excitations in Solids*, Benjamin 1963, chap. 3.

[Pines90] A. J. Millis, H. Monien and D. Pines, Phys. Rev. B42, 167 (1990).

[PinesNoz66] D. Pines and P. Nozières, *The theory of Quantum Liquids* Vol. 1, Benjamin New York (1966).

[PintReich94] L. Pintschovius and W. Reichardt, "Inelastic Neutron Scattering studies of the lattice vibrations of high-T_c compounds", in *Physical Properties of High Temperature Superconductors IV*, Ed. by D. M. Ginsberg, World Scientific Singapore (1994).

[Pist94] F. Pistolesi and G.C. Strinati, Phys. Rev. **B49**, 6356 (1994).

[Pollmann77] J. Pollmann and H. Buttner, Phys. Rev. **B16**, 4480 (1977).

[PV73] K. N. Pathak and P. Vashishta, Phys. Rev. **B7**, 3649 (1973).

[Quattropani79] A. Quattropani, F. Bassani, G. Margaritondo and G. Tivinella, Il Nuovo Cimento B **51**, 335 (1979).

[RaedtLagen84] H. de Raedt and A. Lagendijk, Phys. Rev. **B27**, 6097 (1983); **30**, 1671 (1984).

[Randeria90] M. Randeria, Ji-Min Duan and Lih-Yir Shieh, Phys. Rev. **B41**, 327 (1990).

[Randeria95] M. Randeria, "Crossover from BCS theory to Bose-Einstein Condensation", in *Bose-Einstein Condensation*, Ed. by A. Griffin, D. Snoke and S. Stringari, Cambridge Un. Press (1995).

[RannRobasz85] J. Ranninger and S. Robaszkiewicz, Physica **135**, 468 (1985).

[RannRobin95] J. Ranninger and T. M. Robin, Phys. Rev. Lett. **74**, 4027 (1995).

[Renker88] B. Renker *et al.*, Z. Phys. B-Condensed Matter **73**, 309 (1988).

[Rietveld92] G. Rietveld, N. Y. Chen and D. van der Marel, Phys. Rev. Lett. **69**, 2578 (1992).

[RungeGross84] E. Runge and E. K. U. Gross, Phys Rev. Lett. **52**, 997 (1984).

[Scha55] M.R. Schafroth, Phys. Rev. **100**, 463 (1955).

[Scha57] M.R. Schafroth, S.T. Butler and J.M. Blatt, Helv. Phys. Acta **30**, 93 (1957).

[Schles90] Z. Schlesinger *et al.*, Phys. Rev. Lett. **65**, 801 (1990).

[Schrieffer88] J. R. Schrieffer, X. G. Wen and S. C. Zhang, Phys. Rev. Lett. **60**, 944 (1988).

[Shen93] Z. X. Shen *et al.*, Phys. Rev. Lett. **70**, 1553 (1993).

[ShenDessau95] For a review, see Z. X. Shen and D. S. Dessau, Phys. Rep. **253**, 1 (1995).

[SingTosi81] K. S. Singwi and M. P. Tosi, Solid State Physics **36**, 177 (1981).

[SohFink93] E. Sohmen and J. Fink, Phys. Rev. **B47**, 14532 (1993).

[STLS68] K. S. Singwi, M. P. Tosi, R. H. Land and A. Sjölander, Phys. Rev. **176**, 589 (1968).

[StreetMott75] R. A. Street and N. F. Mott, Phys. Rev. Lett. **35**, 1293 (1975).

[Strinati87] G. C. Strinati, J. Math. Phys. **28**, 981 (1987).

[Stringari92] S. Stringari, Phys. Rev. **B46**, 2974 (1992).

[Sun94] A. G. Sun, D. A. Gajewski, M. B. Maple, R. C. Dynes, Phys Rev. Lett. **72**, 2267 (1994).

[Takahashi89] T. Takahashi *et al.*, Phys. Rev **B39**, 6636 (1989).

[TanTimusk92] D. B. Tanner and T. Timusk, "Optical Properties of High-Temperature Superconductors", in *Physical Properties of High Temperature Superconductors III*, Ed. by D. M. Ginsberg, World Scientific Singapore (1992) and references therein.

[Thomas91] G. Thomas, in *Proceedings of the Scottish Univ. Summer Schools*, St. Andrew 1991, ed. by D. P. Tunstall (1991).

[ThomCard89] C. Thomsen *et al.*, Solid State Commun. **65**, 1139 (1989).

[Thomsen90] C. Thomsen, M. Cardona, B. Friedl, C. O. Rodriguez, I. I. Mazin, O. K. Andersen, Solid State Commun. **75**, 219 (1990).

[Thomsen91] For a review: C. Thomsen, "Light Scattering in High-T$_c$ Superconductors" in *Light Scattering in Solids* Vol. VI, Ed. by M. Cardona and G. Güntherodt, Springer-Verlag, Berlin (1991) and references therein.

[Tosi94] M. P. Tosi, *Introduction to Statistical Mechanics and Thermodynamics* and *Introduction to the Many-body theory*, lectures notes of the courses held at the Scuola Normale Superiore (1994).

[Tralsh91] N. Tralshawala *et al.*, Phys. Rev. **B44**, 12102 (1991).

[Tsuei90] C. C. Tsuei *et al.*, Phys. Rev. Lett. **65**, 2724 (1990).

[Uchida91] S. Uchida *et al.*, Phys. Rev. **B43**, 7942 (1991).

[Uemura91a] Y. J. Uemura *et al.* Phys. Rev. Lett. **66**, 2665 (1991).

[Uemura91b] Y. J. Uemura *et al.*, Nature **352**, 605 (1991).

[Uemura93] Y. J. Uemura *et al.*, Nature **364**, 605 (1993).

[Valles91] J. M. Valles *et al.*, Phys. Rev. **B44**, 11986 (1991).

[vanHarlingen95] For a review, D. J. Van Harlingen, Rev. Mod. Phys. **67**, 515 (1995).

[vanHove54] L. van Hove, Phys. Rev. **95**, 249 (1954).

[Varga65] B. B. Varga, Phys. Rev. **A137**, 1896 (1965).

[Varma87] C.M. Varma, S. Schmitt-Rink and E. Abrahams, Solid State Commun. **62**, 681 (1987).

[VS72] P. Vashishta and K. S. Singwi, Phys. Rev. **B6**, 875 (1972).

[Wachter96] P. Wachter, B. Bucher and R. Pittini, Phys. Rev. **B49**, 13164 (1994).

[Wilczek91] F. Wilczek, Sci. American, 24, May 1991.

[Wollmann93] D. A. Wollmann, D. J. Van Harlingen, W. C. Lee, D. M. Ginsberg and A. J. Leggett, Phys. Rev. Lett. **71**, 2134 (1993).

[Wu90] D. H. Wu and S. Sridhar, Phys. Rev. Lett. **65**, 2074 (1990).

[ZhSato93] H. Zhang, H. Sato Phys. Rev. Lett. **70**, 1697 (1993).

Acknowledgements

I wish to thank my supervisor Prof. Giuseppe Iadonisi for his essential help and advice. I thank Prof. Franco Bassani, Prof. Mario Tosi and Dr. Giuseppe La Rocca for their precious advice in the course of the present thesis.

I would like to address special thanks also to Prof. Sasha Alexandrov for helpful explanations on the small polaron theory and the experimental side of high–temperature superconductors. I would like to thank also all the research staff at the Interdisciplinary Research Centre in Superconductivity in Cambridge and in particular Dr. Nicky Athanassopoulou to have given me her data on London penetration depth. Dr. Paolo Calvani and Dr. Leonardo Degiorgi are gratefully acknowledged for illuminating discussions on infrared experiments on high–temperature superconductors. Thanks are due to my coauthors, Dr. Vittorio Cataudella, Dr. Sergio Conti and Dr. Domenico Ninno.

The European Economical Community, within the Human Capital and Mobility Program under Contract N. ERB–CHRKCT930124, and the Italian National Institute for Condensed Matter Physics (INFM) are gratefully acknowledged for financial support.

Elenco delle Tesi di perfezionamento della Classe di Scienze
pubblicate dall'Anno Accademico 1992/93

HISAO FUJITA YASHIMA, *Equations de Navier–Stokes stochastiques non homogènes et applications*, 1992.

GIORGIO GAMBERINI, *The Minimal supersymmetric standard model and its phenomenological implications*, 1993.

CHIARA DE FABRITIIS, *Actions of Holomorphic Maps on Spaces of Holomorphic Functions*, 1994.

CARLO PETRONIO, *Standard Spines and 3-Manifolds*, 1995.

MARCO MANETTI, *Degenerations of Algebraic Surfaces and Applications to Moduli Problems*, 1995.

ILARIA DAMIANI, *Untwisted Affine Quantum Algebras: the Highest Coefficient of $detH_\eta$ and the Center at Odd Roots of 1*, 1995.

FABRIZIO CEI, *Search for Neutrinos from Stellar Gravitational Collapse with the MACRO Experiment at Gran Sasso*, 1995.

ALEXANDRE SHLAPUNOV, *Green's Integrals and Their Applications to Elliptic Systems*, 1996.

ROBERTO TAURASO,*Periodic Points for Expanding Maps and for Their Extensions*, 1996.

YURI BOZZI, *A study on the activity-dependent expression of neurotrophic factors in the rat visual system*, 1997.

MARIA LUISA CHIOFALO, *Screening effects in bipolaron theory and high–temperature superconductivity*, 1997.

"CompoMat" Loc. Braccone, 02040 Configni (RI), Italy
Finito di stampare nel marzo 1997